SxI – Springer for Innovation /
SxI – Springer per l'Innovazione

Volume 12

For further volumes:
http://www.springer.com/series/10062

Eleonora Riva Sanseverino
Raffaella Riva Sanseverino
Valentina Vaccaro · Gaetano Zizzo
Editors

Smart Rules for Smart Cities

Managing Efficient Cities
in Euro-Mediterranean Countries

 Springer

Editors
Eleonora Riva Sanseverino
Valentina Vaccaro
Gaetano Zizzo
DEIM
University of Palermo
Palermo
Italy

Raffaella Riva Sanseverino
Department DARCH
University of Palermo
Palermo
Italy

ISSN 2239-2688   ISSN 2239-2696   (electronic)
ISBN 978-3-319-06421-5   ISBN 978-3-319-06422-2   (eBook)
DOI 10.1007/978-3-319-06422-2
Springer Cham Heidelberg New York Dordrecht London

Library of Congress Control Number: 2014943509

Printed on acid-free paper

Springer is part of Springer Science+Business Media (www.springer.com)

Preface

Modern cities require new and different codes, taking into account environmental and energy issues, pillars of contemporary urban development. The existing urban communities are highly energy consuming systems, strongly contributing to greenhouse gas emissions. Contemporary cities cannot be thought of and defined as static systems, as they were in the past, with a few urban functions. "The Church, the square and the court, the marketplace and the government palace were in the past the places where the urban functions were taking place" (Salzano 1998). Today, the reference frame has changed even due to the demographic increase more and more concentrated in the cities that are now large concentrations of streams of people with new needs.

New parameters must now be considered together to plan how to reach the desired urban *smartness* (energy, mobility, waste...).

The urban mobility, as an example, has a growing importance, since it directly and locally affects the quality of air; such as the issue of energy production, that strongly influences the planning approaches, and forces the cities administrations to guess new paths for the future development of urban settlements. Some cities in the world are getting closer to a new urban model (*smart* cities), where planning the urban functions must be carried out in an integrated manner.

In the Arabian desert new "zero emissions" cities are being built (Masdar) and in the old Europe the energy planning in cities is becoming a basic element of the urban development, with an increasing number of cities that envision their development while reducing the CO_2 emissions, increasing the energy efficiency and the production of energy from Renewable Energy Sources, as required by the EU in the "European Union climate and energy package 20-20-20."

The scenarios delivered by cities are not always comparable and it is difficult for highly disadvantaged settlements to start a virtuous process especially in this moment of economic crisis. In Italy, some cities are setting sustainable actions for the development of the urban organization—green economy, sustainable mobility for the limitation of greenhouse gas emissions—while other cities in Italy as in other countries are not able to devise plans and, overwhelmed by emergencies, keep on thinking of city development in the traditional way.

Among the European cities some leading examples can be found. Copenhagen is the new Green Capital of Europe (2014); it is a city that has balanced development and innovation, such as Stockholm (European Green Capital, 2010) and Amburg (European Green Capital, 2011).

The lesson learned from these examples is that cities should promote innovative mechanisms to design and manage the processes in order to improve the quality of living in the urban areas.

New rules and new codes are needed and new integrated functions and responsibilities in the city administration must be considered. Integrated plans must be devised for cities, where sectorial planning is disappearing to leave the floor to inter-disciplinary design.

Moreover, from the analysis of what exists, the cities of the future seem to be *smart* if their development plan follows the urban supply chain following a *bottom-up* approach. Energy is one of the most important elements of such chain, since it interacts with the production and consumption mechanisms of the city, influencing the choices of urban and buildings design. But not only. Soil, water, and materials consumption as well as recycling are crucial for the identification of the new functions of the city.

New rules will define the urban and building performance requirements unifying the design and building processes of cities (acoustic pollution controldir. 89/106/CEE; solar energy and energy saving dir. 89/106/CEE, 2002/91/CE, 2006/32/CE) considering also the other functions. A *bottom-up* approach to support a smart development of cities is the urban and building code.

At different scales, such rules will allow the definition of the not yet so clear boundaries between uses and functions, public and private, inherited by the contemporary urban vision. The integration of functions and the sharing at different levels in the use of urban spaces and services (public and private) should define the urban quality, in terms of measurable performance indices. Proposals and measures have been indeed till now separately formulated in different fields.

In the use of energy field, the project Address in the VII Framework Program, now at its end, has put together the largest energy distributors in Europe and has ideated a new actor of the energy market: the aggregator. The latter interfaces the small private customers with the distribution companies, creating a shared demand profile that allows to reach energy efficiency objectives for the electrical network. In the same way, in the field of urban mobility, car sharing and bike sharing are ways to share a collective transportation service, such as the use of recharging systems for electric vehicles. These systems if adequately controlled may offer regulation services to the energy distribution system (vehicle-to-grid, V2G), putting intelligently together, again, the private resources of many users.

All this motivates and gives substance to the new urban design of uses and functions for different services offered to citizens through Information and Communication Technology (ICT).

This design level allows the definition and management of the "active" urban quality. Other measures, concerning structural features, influence the so-called

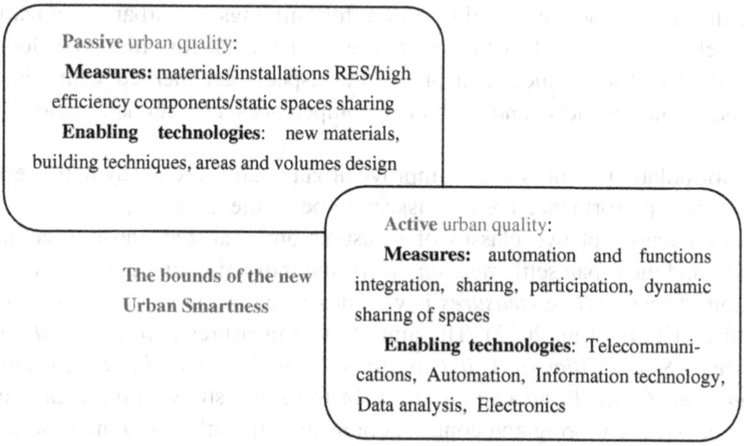

Passive urban quality:
Measures: materials/installations RES/high efficiency components/static spaces sharing
Enabling technologies: new materials, building techniques, areas and volumes design

The bounds of the new Urban Smartness

Active urban quality:
Measures: automation and functions integration, sharing, participation, dynamic sharing of spaces
Enabling technologies: Telecommunications, Automation, Information technology, Data analysis, Electronics

Fig. 1 The bounds of the new Urban Smartness

"passive" urban quality (materials, installations of RES based generation, etc., namely nonautomated components) (Fig. 1).

In this scenario, Italy, as other countries, is going through a deep change at local administration level: the recent experiences confirm the importance of understanding what is happening in contemporary society more and more culturally oriented towards "sharing" through the new media and what the technical and industrial communities can offer as a solution to cope with the new requirements of sustainability.

In this manuscript, a methodological approach to devise a new and *smart* urban/ building code for local administrations is proposed. The study analyzes what exists and tries to set out a methodology, taking into account the regulatory framework and the economic feasibility of the proposed measures taking as case study a large Mediterranean city in Italy. Such economic feasibility is evaluated using the existing European regulatory framework (directives and technical norms). In this study, Italy is taken as a paradigm of the evolution of modern cities between historical heritage and bureaucracy.

The perspective in writing this contribution is that of a city overlooking the Mediterranean Sea in Italy, where cultural obstacles are posed by individualism at all levels against the concept of sharing resources and of attaining common goals.

Nonetheless, many local administrations in Italy have indeed already taken some steps in the direction of modifying *towards sustainability* the existing building and urban codes. It is interesting to underline how in Italy, year by year, is growing, not only the number of local administrations (+42,3 % as compared to 2010 and +80 % as compared to 2009), but also the number of different relevant issues dealt with. And nowadays, such new *smart* building and urban codes are formulated and adopted in almost all the areas of Italy, with at least one innovative building or urban code per Italian region.

The interest around a new ruling system for buildings and urban organization at local level resides in the fact that codes are fundamental for cities' development, since within codes technical and procedural aspects get merged with economic interests, social impacts, and technical competences at national, regional, and municipal levels.

The formulation of measures to improve urban quality and a way to numerically evaluate their performance are also issues to be studied.

In this manuscript two classes of measures are analyzed, those affecting the buildings and the urban settlement *passively* and those affecting them *actively*. The definition of these *active measures* is well described in the work "Smart cities of the future" (Batty et al. 2012). The aim of such measures is *to Relate the Infrastructure of Smart Cities to their Operational functioning and planning Through Management, Control and Optimisation*. In the same study, some technological tools to devise monitoring and control actions over the urban system, described as a network of services, are analyzed. Such active and passive measures are already considered in the European regulatory framework and should be organically integrated into the local building/urban codes according to the local specificity of climate, society, and existing national regulatory framework.

Since 10 years, the European Union is one of the main actors in this path to sustainable development of cities with *bottom-up* measures for the building sector, with the aim of increasing the contribution of the building sector to greenhouse gas emissions reduction and to RES-based generation support. In this context, the challenge of European directives is that to fix for the member states objectives and methodological approaches through suitable Technical norms. As prescribed by the directive 2010/31/UE (the new EPBD, Energy Performance in Buildings Directive, also called EPBD recast), since January 1, 2021, it will be only possible to build neutral buildings in terms of Energy consumption, called Near Zero Energy Buildings (NZEB), showing a very high energy performance. The nearly zero or very low amount of energy required should to a very significant extent be covered by energy from renewable sources, including renewable energy produced on-site or nearby. Another important change brought by the directive concerns the objectives proposed, that are now defined in terms of numerical performances. This means that, in cities, buildings will have to be thought of, designed, and built to reach precise and measurable objectives in terms of energy efficiency for heating and cooling, that can be reached using different available measures (active or passive) that best appraise the RES-based supply within the considered urban context. The change is enormous as compared to the habit of many existing local administrations, for which the gap filling will require educational support and an attentive communication strategy, as well as an adequate coding system, to accompany the growth of new competences, the experimentation, and the definition of protocols and rules (Report ON-RE 2012).

In the design of new coding systems, it is fundamental the economic feasibility of measures and the cost-benefits analysis to which the codes refer.

Such evaluation is particularly interesting when the increase of energy efficiency is attained through Building Automation technology. The recent directive 2012/27/ UE has been issued as a compromise between the business world producing components for Building Automation and the public administrations structurally in economic shortage. This Directive establishes a common framework of measures for the promotion of energy efficiency within the Union in order to ensure the achievement of the Union's 2020, 20 % headline target on energy efficiency and to pave the way for further energy efficiency improvements beyond that date. According to the cited directive, the member states must elaborate an *action plan* with precise commitments in relation to the objectives of the directive. Each member state guarantees that since January 1, 2014, the 3 % of the covered surface of the heated/cooled public buildings, except the already cited NZEB, is restruc- tured every year to reach the numerical requisites of energy efficiency imposed by the directive 2010/31/UE.

Notwithstanding the constraints set forth by the limitation to the public indebtedness (further to the principles of "stable prices, sound public finances and monetary conditions and a sustainable balance of payments" of Article 119 of the Treaty on European Union and the Treaty on the Functioning of the European Union) and the financial and economic crisis, the Italian legislator has imple- mented some measures to support public and private entities in the development of smart cities, including the smart (i.e., efficient) use of energy. However, nowadays, it is still difficult to turn energy efficiency actions, which at the end entails a saving, into cash flows, which can be financed and can guarantee the loans granted by the lenders. Furthermore, the credit crunch and the lack of knowledge on the side of the public and private entities (who should promote the energy efficiency interventions) and on the side of the lenders hinder the funding of energy efficiency initiatives.

References

Batty M, Axhausen M, Giannotti F, Pozdnoukhov A, Bazzani A, Wachowicz M, Ouzounis G, and Portugali Y (2012) Smart cities of the future. Eur Phys J 214(1):481–518

Salzano E (1998) Fondamenti di urbanistica. Laterza editions, Roma-Bari

Contents

1   **Competitive Urban Models**   1
Raffaella Riva Sanseverino

2   **Smart Cities Atlas**   15
Raffaella Riva Sanseverino

3   **The Integration and Sharing of Resources for a New Quality
of Living**   29
Raffaella Riva Sanseverino and Salvatore Orlando

4   **Urban Smartness: Tools and Experiences**   45
Domenico Costantino

5   **The Urban and Environmental Building Code
as Implementation Tool**   59
Valentina Vaccaro

6   **Economic Feasibility of Measures for Energy Efficiency**   87
Eleonora Riva Sanseverino and Gaetano Zizzo

7   **Funding Energy Efficiency Measures**   109
Silvia Dell'Atti

8   **Smart Planning and Intelligent Cities:
A New Cambrian Explosion**   123
Maurizio Carta

Competitive Urban Models

1

Raffaella Riva Sanseverino

Abstract

The smart city model is here described according to the most recent studies on the topic. Indices and numerical features concretely evaluate the different aspects of this *model* in two analyses whose results are reported and commented in this chapter.

1.1 Smart City Model

Technological and inter-connected, but also sustainable, comfortable, attractive, safe, in one word "smart": this is how the smart city looks like, a model of city on which, in Europe and in the world, governments are betting to provide a balanced urban development keeping up with the demand of welfare, coming from the middle class.

Aiming at technological innovation to improve management of urban processes and quality of life of citizens, this is the direction followed by some local administrations in Europe, that are starting projects, and setting agreements to re-draw cities. In relation to the objectives fixed by the EU, supported by 'pacts' and formal 'commitments', all cities are involved in this transformation process that should turn them in different ways *in smart cities*.

R. Riva Sanseverino (✉)
Department DARCH, University of Palermo,
Viale delle Scienze, 90128 Palermo, Italy
e-mail: raffaella.rivasanseverino@unipa.it

E. Riva Sanseverino et al. (eds.), *Smart Rules for Smart Cities*, Sxi 12,
DOI: 10.1007/978-3-319-06422-2_1,
© Springer International Publishing Switzerland 2014

The expression *smart city* is spreading more and more and takes progressively a precise meaning after the recent results carried out in the experimental phase in different urban contexts.

Over and mis-used expression, it implies a really silent revolution in our cities, often sat back in wrong habits and not attentive to the cautious management of resources and consumptions.

On the Internet everything becomes suddenly *smart*, cars (the 'smart' city car produced since more than ten years) as well as consumer goods and the adjective typically recalls something new, easily marketable and *cool*.

But what is really a "*smart city*"? It is a city that bets a lot on the quality of living and where the citizens are involved as main actors in decision processes (Dominici 2012). At the basis of creating a smart city there is certainly a new and integrated design process, aiming at a new modulation of the urban functions (both the traditional and the new ones appearing in everyday life) also thanks to the digital technology innovation.

In this way, the traffic can be monitored and influenced through smart phones able to identify the congestion areas and to set alternative routes for city trips: a true digital traffic officer able to let people move in cities without delays.

The basic condition of such scenario is the existence of adequate infrastructures able to support such innovations (the design of streets, the presence of an efficient public transportation network and cycling routes, ICT facilities).

A smart city is thus a city that does not pollute for standard functions deployment and extensively uses renewable energy sources.

As we will see, indeed, three are the main features of smart cities: smart mobility, smart energy and smart governance, where the participation and sharing of opinions takes place. The latter aspect is fundamental to grow public acceptance of sustainable ways of living the cities. Sharing innovative processes is indeed a fundamental aspect that cities have set out in different ways. The city of Bari, in Italy, as an example has ideated a no profit association named Bari Smart City[1], putting together public and private entities with the aim of creating a productive environment where it is possible to design, together with the representatives of the society and of the business world, solutions and ideas for a sustainable urban development model. The associative model should serve as a way to understand how innovation may re-design the ways of living of citizens.

In these times, an obstacle certainly is the lack of economic resources required to implement the measures: the municipal administrations indeed, constrained by everyday emergencies are forced to devise new tools such as the activation of public-private partnerships to access the financing calls (such as for example the European Jessica[2] funding scheme, etc....).

---

[1] The following sites show the development of Bari: www.barismartcity.it/; www. osservatoriosmartcity.it/bari/; www.greencityenergy.it/bari/.

[2] JESSICA (Joint European Support for Sustainable Investment in City Areas - Joint European Support for Sustainable Investment in City Areas) is an initiative of the European Commission in collaboration with the European Investment Bank (EIB) and the Development Bank of the

Smart Energy is another important declination of the intelligent city: for this reason, in Italy cities like Genoa and Turin have mostly invested on smart electrical grids and Renewable Energy Sources, RES.

The first objective to be reached to create an intelligent city is indeed an optimized system for the management of Energy resources that are able to put in motion most of the urban functions. Both public and private transportation can be managed efficiently as well as waste and water, and of course heat, electrical energy and data.

Such optimized system must be in most cases, due to the technical features of the functions to be implemented, based on ICT. New technologies are indeed the further step to be taken to improve the management of urban processes and as a result the quality of living of citizens.

The process leading a city in Europe to become *smart* is accompanied by some important steps such as undersigning *the Covenant of Mayors*[3] and designing an *Action Plan for Sustainable Energy*[4]: the latter contains the strategic measures that must be set out to get certain objectives for sustainable development.

1.2 European Smart Cities

The most articulated investigation concerning the smart city is certainly the one elaborated by the University of technology of Vienna (Giffinger et al. 2008) in cooperation with other European research centres, over a sample of 70 middle size European cities. The researchers of the technical university, together with the university of Lubiana and the poli-technical university of Delft, have carried out since 2007 a research about the European Smart Cities, developing a ranking tool specific for cities with a population below 500.000 inhabitants. The results of the study, has made a comparison among 70 European cities. The research has started from the premise that around 40 % of the entire European population lives in small cities and most of these cities, even though they show a big potential, often stay in the shade as compared to the larger European cities. The competitive advantage small cities have and that is often not considered by investors is their limited size, both in terms of population and in terms of territorial extension, which allows to easily test new solutions.

(Footnote 2 continued)
Council of 'Europe (CEB). It promotes sustainable urban development and urban regeneration through financial engineering mechanisms.

[3] www.eumayors.eu/.

[4] A Sustainable Energy Action Plan (SEAP) is the key document in which the signatory Administration outlines how it intends to reach its CO_2 reduction target by 2020. It defines the activities and measures set up to achieve the targets, together with time frames and assigned responsibilities. Covenant signatories are free to choose the format of their SEAP, as long as it is in line with the general principles set out in the Covenant SEAP guidelines.

According to the system proposed by the researchers of the Technical University of Vienna, the ranking tool proposed to identify the smartness level of European cities is declined in six specific features: economy, people, governance, mobility, environment and living.

Seventy test cities have been selected in Europe showing the following features:
- urban population between 100,000 and 500,000 inhabitants;
- basin of users and workers lower than 1,500,000 inhabitants;
- at least one university.

To compare and rank them in terms of *smartness*, it has been necessary to standardize the measurement criteria of the 6 features above, obtaining for each of them, a numerical evaluation to be given to each city.

Among the 70 cities also 4 county seats of Italian cities are present. The best placement was obtained by the city of Trento, historically on top of life quality in Italy, at the 45th absolute position, followed closely by Trieste, at the 49th absolute position, Ancona and Perugia, respectively can be found at the 51st and 52nd absolute positions.

1.2.1 Efficientcities: Cities and infrastructures for growth in Italy

It is on the field of urban mobility that the local administrations in Italy are investing more to implement the concept of smart cities in Italy. This is what emerges from the recent study (Efficientcities Siemens 2012) "Efficientcities: cities and infrastructures for the growth of the country" carried out by a public consortium supported by municipalities, Cittalia-Anci research, and Siemens Italia that analyses 54 main Italian cities with more than 90 thousand inhabitants, whose performances, in terms of innovation and quality of living, have been analysed using a set of indicators for different areas (green, water, air, waste, building heritage and quality of living, energy, health, mobility and logistics).

The study shows that urban mobility is the most financially supported area with 10.7 billions euros, in three years, followed by the sustainability of buildings and water.

The other investment fields are: public and private building heritage and urban restoration respectively with 2.4 and 2.1 billions euros of forecasted investments. The overall value of the three years plans for the 54 cities is around 37.7 billion euros. This value, if compared with the Italian GNP 2011 (1.579,7 billion euro) leads to the conclusion that the overall amount of planned investments, if actually made, would be around 2.39 % of the Italian GNP.

The study classifies the cities considering 6 different performance aspects (quality of the urban environment, welfare cities, ideal cities, quality of living and mobility, cities in transformation and energy cities). The objective of the study is to identify the basic infrastructural requirements for each of the aspects above, through the definition of homogeneous classes of cities and identifying within each class a model representing at its best the features of the group to which it belongs (benchmark model).

From the point of view of quality of the urban environment (including management of building heritage, the waste handling and the cycle of water), the study puts into evidence how some southern cities in the Puglia region (Foggia, Andria, Barletta and Lecce) are among the first places. As far as quality of living and especially mobility is concerned, a clear geographic diversity emerges: all the northern or centre Italy cities are in the highest rankings, while the southern cities show a very low position in the overall ranking. Milan, with its optimum performance, is only behind Bergamo. Going to renewable energy use (cluster energy cities) Forlì, Trento and Ravenna have the highest positions. More in general, again, the cities in northern Italy show improved performances as compared to the Southern cities. Positive exceptions in the South can again be found in the region Puglia, especially Foggia and Lecce. In the clusters identified by the study Siemens-Cittalia, Reggio Emilia is the benchmark city for the urban environment. The relevant cluster, composed of 17 cities all at the Center-North of Italy (except Sassari in Sardinia), has gathered cities that differ from the others in terms of entity of the investments for territorial management and production of energy from renewable sources. Among the welfare cities Cagliari emerges in a group that puts together, among the others, important coastal cities (Bari, Genoa, and Naples), showing an important building heritage and an health care system that attract people from all over the country. In the cluster of ideal cities, where the quality of living for citizens is outstanding, Trento emerges among the centre-northern cities (Bergamo, Brescia, Padova and Trento); while in the cluster of quality of living and mobility Venice is the leading reality in a group that collects, except from Bolzano, almost all the large cities in Italy (Bologna, Firenze, Milano, Roma, Torino). Below average in most aspects are mostly southern cities and in particular 10 cities: Palermo, Messina, Catania, Reggio Calabria, Catanzaro, Barletta and Pescara, but also other cities located in the centre and in the north like Pistoia, La Spezia and Trieste. Nonetheless the latter cities are described as having a strong potential for improvement, such as Pescara, especially in terms of health and environment.

Finally, the 8 cities belonging to the cluster energy are quite good in terms of clean Energy production systems but show values below the average for most of the other indicators. The best performance in this cluster are provided by the city of Lecce, benchmark city for this group in which also other cities from the Puglia region can be found (Andria, Foggia, Lecce and Taranto), from Sicily (Siracusa) and from the centre of Italy (Arezzo, Terni and Latina). The just seen approaches for smart cities analysis at European level and at Italian level are very different due to the different in-depth analysis carried out. The different width of the scenario, Europe in one case and the Italian country in the other, and the different times in which the studies have been carried out produce very different results. The European study carried out in 2007 on 70 small cities, hosting a high level education centre, could make use of limited previous studies on the topic while the Italian study (completed in 2012) has been based on larger competences acquired in more recent times. Moreover, due to the particular historical context of the Italian country the Efficient cities study is tailored on smart indicators that in other countries may just be inapplicable, such as the attention to the building historical heritage and its management.

1.3 The Key Features of the Smart City: Smart Governance, Smart Mobility and Smart Energy

1.3.1 From Traditional Governance to Smart Governance of Cities

The last years have seen an increasing complexity of the political and administrative situation of countries with a consequent reduction of the possibility to implement adequate measures on the territories. The economic crisis signs with absolute evidence the current times.

The communication between public administrations and citizens has dramatically changed with the advent of new information technologies and the Internet.

The recent political elections at National level in Italy—with a success of the *Movimento 5 Stelle* of Beppe Grillo irradiating its principles from the Internet— have shown the enormous communication potential of the network. In the same way, the *Pirates* party in Germany describes itself as part of an international movement supporting the "digital revolution" as a means of increasing transparency and participation in politics.

The possibility to spread in a wide space and in real time every possible form of expression that can be digitalized has created innovative processes also in the National Public Administrations, centered on efficiency, effectiveness, and transparency and allows saving money in the *Res Pubblica* administration. At national level, in Italy, the public administration is composed of different subjects of different sizes and needs (mainly Regions and municipalities). All these subjects, but in particular the municipalities, have some common features such as the need to design and implement optimization processes for their resources, including the information technology infrastructures, so as to attain strong improvements of the quality of the services offered to citizens without increasing the level of current expenses.

In the aim of attaining a full development of the digital revolution, not only the organizational aspects of the public administration must be considered, but also it is necessary to implement a motivational and educational path for those who work inside public administrations.

Italy is certainly below the average level of literacy rate in English and in Information Technology in public administrations; this is caused by the lack of motivation by public employees, hardly appraised by merit and suffering from the lack of infrastructures.

Although it is now mandatory, the Italian municipalities (Riva Sanseverino 2013) still do not even try to apply for European financing calls, due to the abovementioned reasons. Another relevant aspect in a smart governance system is the extent to which the community is interested in taking part to the decision process. In the past, the participation level was very low and was limited to a small oligarchy; more recently, the spread of television programs based on talk shows has increased the level of consciousness in people, but interactivity was not allowed. Now interactivity is the new vehicle through which information is delivered, at all levels, and it seems that also politics and common goods can be managed using the citizens opinions travelling on the Internet.

This new concept of Smart Governance connects the technical aspects to the social ones meant as not only a strictly governance tool, but also as an educational and communication tool for thoughts and opinions of the citizens.

With the expression *Social Media* are to be intended tools and services created in contrast to those that are defined industrial media, such as newspapers, radio, TV and cinema.

The classical media have one direction transmission and centralized creation of contents; a few people create the information and decide what to transmit, and a lot of people are passive users of this service. The Social Media are at the exact opposite, allowing a lot of people to create the contents and to spread them.

This type of participative media can be implemented thanks to the technology. If with the first generation Web only tentative forms of social participation were introduced, with Web 2.0 the Social Media are fully implemented and the so-called User Generated Content (UGC) characterizes them.

This fundamental transformation on social participation has shifted the centre of gravity on which marketing has its foundations for years. In traditional media the message had to be supported on a one-direction channel with no possibility of contradiction. In *social media* it is instead quite possible that the consumer expresses and opinion, and in general, his contribution is fundamental to support the spread information about that item among a lot of people.

The proliferation of such environments for social media is huge, but they can be divided into two main categories: Blogs and Social Networks. Blogs are virtual environments where individuals, groups of people or enterprises publish contents that are considered interesting for the community. The subjects can be divided in classes or general, but most important is the possibility given to users to take freely part to discussions that arise on the different subjects. All this allows generating what is called counter-information, and that is basically important for all those who want to 'understand' the market.

The Social Networks are instead virtual environments finalized at interconnecting people based on features or common interests, everything in the aim of sharing resources or simply discuss of any kind of subject.

Worldwide phenomena like YouTube, Facebook, and Tweeter are some of these platforms where people can share something. Photos, videos or simply written text, what brings together these Social Networks is the enormous participation of people. This obviously makes imperative the consideration of these platforms as the most important places where desires and people's needs can be figured out.

The technology supporting these systems can also be employed to influence political decisions at all levels, including those concerning the urban environment and thus uses and functions in cities, buildings, urban spaces, energy, waste, water, living, etc. In this way, ideally the democratic process can be enhanced and public services can better address the needs of citizens.

The objectives of this new communication system are the stimulation of ideas to encourage and support the exchange of experiences and generate new methods of working. A virtual network with an efficient and trim administrative structure, that carries out the events management through opinions sharing and regular updates.

The *digital revolution* in politics called *E-government* (Profiti 2011) is already changing the way in which politics interacts with citizens and commercial entities and the *Smart Governance* system puts a challenging and higher objective for our administrators.

The real main actor of the process is however the citizen that has to lead the era of urban transformation, and that is now able to express his opinion on any kind of issue affecting his everyday life.

The ultimate goal of such technological evolution is that to turn the governmental functions from an 'office-centric' working mode to a 'citizen-centric' working mode.

E-Government is intended as the use of ICT in public administrations, accompanied by a change in the organization and by the acquisition of new competences from the employees, with the aim of improving the services offered and supports the democratic process and welfare.

An efficient planning tool for the urban context and buildings must also account for the communication infrastructure that puts in synergy multiple aspects of the city and allows the coordination of the different functions.

ICT is thus the enabling technology through which privates, administrators and business people meet influencing the administrative actions; in this way it is possible to support the participation of citizens to the decision process, increasing the sense of belonging of the individuals to the community.

1.3.2 Intelligent Urban Mobility

Our cities currently show a mobility system for people and goods that are not sustainable; indeed many researches and analysis have evidenced that the common use of the car generates congestion, pollution and high CO_2 emissions.

To limit these unfavourable effects of the use of car, it is required to change the vision of cities, limiting the use of private cars, supporting the use of collective means, the sharing of the service offered by cars instead of keeping the property for single individuals, the use of bikes, the inter-modality of transportation services.

Also the technology for building cars is radically modified: small and light vehicles with zero or reduced emissions.

In this new scenario, freedom of moving in cities goes necessarily through a radical change of habits in cities. The motivations underlying the massive use of private cars, often resides in missing or not efficient public transportation services.

Besides, in most cases, especially in Italy, the first problem to be faced is that the recently urbanized settlements, both concentrated and diffused, have been built precisely because individuals own private means of transportation.

The second problem arising is one of the main factors conditioning the environmental pollution and namely the Internal Combustion Engines technology for cars.

The third problem is that the new settlements (far from the centre, with a large extension and at low density) are designed and realized (at least in Italy) despite the existence of a shared transportation system (tram, metro).

Those who have invested indeed, have only aimed at the gain deriving from the change of destination of the soil with no refrain from the public administrators and the technicians.

The fourth problem is that the current life style implies an increasing frequency of trips in the cities also increasing the mobility of people and goods.

Smart mobility in a city is thus:

- the ability to guarantee a good availability of public, innovative and sustainable transportation services;
- the support of low environmental impact transportation means such as bikes or pedestrian routes;
- ruling the access to historical centres.

Smart mobility also means the adoption of advanced solutions for the mobility management through info-mobility, managing the mobility of individuals within the city and towards the neighbouring areas.

Smart mobility is thus intermodal, optimized and efficient and allows minors to move independently at no risk not demonising the use of private cars.

In some cases, traffic limitations have been imposed through pedestrianisation and urban tolls. These actions have produced valuable and interesting results, but according to some observers the risk is to create a mosaic of islanded urban areas.

Some local administrations, as an example, limit the access to the city centres based on the *norm Euro* classification, by which the technical requirements for the homologation of the vehicles of motor concerning the emissions have been settled down, to avoid that they defer from a Member State of European Union to another one.

Other administrations use different criteria. Many actors desire orientations and harmonized norms at European level for "clean urban areas" (pedestrian islands, limited access, speed limits, urban tolls etc.), so as to support the spread of such measures all around Europe in and uniform way, allowing an harmonized development of cities. Moreover the harmonization and inter-operability of enabling technologies will allow the costs reduction.

Other measures to improve sustainability of transportation within cities are the implementation of inter-modal mobility as well as *carpooling* and *car sharing* services, the introduction of the "Mobility Manager" and the support to the so-called *soft mobility* (on foot and bikes). In this context, also the adoption of incentives for alternative fuel based public or private transportation means, using RES and not competing with the food sector and thus really sustainable, is advisable.

The recently started European project *MobiEurope*[5], promoting the electric transportation in cities supported by a strong ICT infrastructure, among the expected results reports around 30 % of the total energy consumed for e-mobility deriving from Renewable Energy Sources and mainly wind power.

The ultimate aim of the project is to bring together four major on-going electro-mobility initiatives, effectively contributing to a pan-European interoperable and integrated smart-connected electro-mobility.

The idea of a EU register office for vehicles, with trans-boundary validity all over Europe, suggested by some actors is certainly an interesting proposal, which deserves further study. The elaboration of the data about traffic and vehicles paths can give indications, assistance and dynamic control of transportation to passengers, drivers, and operators in the field of goods and people mobility all across Europe.

Some applications are already available for highways, railways and river transportation. In the next years, these applications will be strengthened by the satellite system Galileo[6], which includes at most 30 satellites in orbit at around 20,000 km of height and that will allow a really precise localization of transportation means.

The current organization of transportations within most cities in Italy for example is characterized by a predominance of traffic on the roads and by the use of private vehicles. This aspect, else than representing one of the most significant causes of green house gas air pollution, jeopardizing the health of citizens, influences the efficiency of the urban areas.

As a conclusion, sustainable mobility for cities is an urban mobility system able to put together the need for reducing the so-called negative externalities connected to it and efficiency.

Each measure to get these objectives has pros and cons and different impacts considering the local features at social, regulatory and economic level and represents also political views about the development of cities.

---

[5] MOBI.Europe aims to make users more comfortable with the use of EVs beyond the limits of "range anxiety" by providing them universal access to an interoperable charging infrastructure, independently from their energy utility and region.—See more at: http://www.mobieurope.eu/the-project/objectives/.

[6] European system of satellite location, competitor of the american GPS, operational since the beginning of year 2014.

1.3.3 Smart Energy in Europe: Comparing Sweden and Italy

The energy sector (Riva Sanseverino et al. 2014) is certainly a strategic field for countries and contemporaneous cities: all essential urban functions are indeed supplied by energy.

Comparing southern and northern Europe cities is also a challenging task due to the different contexts in which such cities have developed within the main European frame.

The commitments taken by the EU in 2008 with the climate package 20-20-20 have motivated all European cities in setting specific actions to reach the objectives of energy efficiency and low carbon energy production for the reduction of GHG emissions.

Energy production and use therefore are the pillars of the above package of directives. And the relevant specific objectives have been then turned into specific national objectives. For Italy, as an example, as far as clean energy production is concerned, the objective is to increase up to 17 % the coverage of the energy required for consumption from renewable energy sources by 2020 (in 2005 it was 5.2 %).

Some European cities supported by serious national policies are strategically directed towards a smart approach in the energy sector, considered of basic importance for the urban development and for the political independence of the entire state they belong to. The European Environment Agency (EEA) estimates indeed that in some cases cities and towns account for just 69 % of national energy use. This is achieved in a range of ways, from increased use of public mobility due to larger population density to smaller city dwellings that require less heating and lighting.

In Sweden[7] the national goals are clear to all operators since years. The production of energy in Stockholm (Nylund et al. 2010; Stockholm City Council 2010a; Stockholm city Council 2010b), as an example aims at self-sufficiency as well as at reduced environmental impact using different means: from the well known renewable sources installation to innovative systems allowing—especially when building new districts—the re-cycling and re-use of waste, the biogas production or even the use, as in the building Kungbrohuset (2010–2011) in Stockholm, of the heat produced by humans in the underground for air pre-heating.

The situation in the Swedish country besides is totally different from Italy from many points of view. In Sweden, the project for the community comes first as well as the carbon-free and politically independent vision of future scenarios that all parties recognize to be a priority. In Italy, the individuals come first and before the interests of the community. What indeed is amazing is that while Sweden struggles since many years to attain the cited objectives of becoming a carbon free society (district heating started in the seventies); in Italy this is a recently debated topic.

---

[7] Interview to M. Ermann, Strategic Planner, Municipality of Stockholm (Riva Sanseverino et al. 2012).

The strategies put in action by Sweden are 360° and aim at reaching the ambitious goal for the city of Stockholm to be a totally fossil free city by 2050.

In this way, the energy plan of 2008 is subordinate to the Regulatory Plan (Costantino and Riva Sanseverino 2012) in agreement with the decisions of the Municipality of Stockholm taken in March 2010. Besides, a major cooperation exists between Municipalities and energy distributors to find the best installation areas for components and plants for the distribution system; them both supporting the use of remote heating systems and the use of renewable sources for private use.

The production of energy in Italy instead is still strongly constrained by the use of fossil–fuel based generation systems.

The choices of energy policy in Italy have been strongly biased by the existence, since many years, of public monopoles in the energy field. The main fossil fuels extraction and distribution company ENI and the Italian power distribution company ENEL are still controlled by the Italian Ministries of Economy and Finance. The slow pace at which all political decisions go in Italy due to a dramatic mix of bureaucracy and individualism puts constraints over the possibilities to modify a route, although recently ENEL has been investing in new and greener technologies (ENEL green power) and has been on the forefront for remote measuring systems production and installations.

As in Sweden, also in Italy, reaching independency in the energy sector would be a desirable goal. Italy depends on imports for most of its fossil fuels. It now produces only 10 % of the natural gas it consumes, compared with 90 % in the early 1970s. Production has fallen steadily while demand has increased, driven largely by growing demand for electric power: 40 % of Italy's natural gas consumption is now for power generation.

Russia and Algeria are currently supplying two thirds of Italy's gas needs, through pipelines via Austria and Tunisia-Sicily. In 2009 natural gas was responsible for over half of electric power generation (60 %), coal 12 % and oil 8 %, the renewables sources including hydropower supplying the remaining 20 %.

The dependency from other countries has produced in Italy quite high energy prices in Europe and the third highest increase of the energy price in the industrial sector in the years 2011–2012, above 15 %.

The reason of this is two fold: the dependency from other countries and the mix of primary sources for producing energy in Italy, which is unique in the European scenario.

Following a 1987 referendum Italy banned nuclear power, but the government reversed this decision in 2008. Recently, following the events at Fukushima, a moratorium on nuclear suspended new nuclear developments. As a result, electric power prices are high in Italy.

Besides such prices also account for the incentives provided for photovoltaic installations support (feed in tariffs) in these years and taxes. This means that the more expensive technologies for renewable electricity can easily attain the sought-after 'grid parity', more easily than in many other markets—and this is already close to the truth for solar photovoltaic installations in southern Italy.

Interestingly, in Italy, despite the unclear reference framework, a smart 'bottom-up' process involving the energy sector has started right from cities. Cities from all over the country, from south to north, show their willingness to reduce their environmental impact. Cities from the centre and from southern Italy (Lecce, Taranto, Andria and Foggia in the Puglia region; and Arezzo in Tuscany, Terni in Umbria, Latina in Lazio and Siracusa in Sicily), but also from northern Italy, such as Modena in Emilia Romagna are starting to put money into new infrastructure for efficient energy production and use.

An interesting case is the city of Modena, whose council joined the Covenant of Mayors[8] in January 2010 and developed a Sustainable Energy Action Plan for the Municipality of Modena with a financial support from the project come2CoM (Intelligent Energy in Europe). The document was taking into account the inputs from public and private stakeholders, to reach the goal of the Covenant of Mayors to reduce the CO_2 emissions of 20 % by 2020.

Municipalities are typically strongly constrained by small financial resources. To turn the Covenant of Mayors commitment from targets into concrete actions, the Municipality (like Modena) has to look for alternative financing solutions.

Energy Performance Contracting (EPC) offers the Municipalities a solution to start on the road to achieving 2020 goals. An EPC is a performance-based procurement method and financial mechanism for building renewal. Utility bill savings, resulting from the installation of new building systems reducing energy use, pay for the cost of the building renewal project. A "Guaranteed Energy Savings" Performance Contract obligates the contractor, a qualified Energy Services Company (ESCO), to pay the difference if at any time the savings fall short of the guarantee[9].

Prompt financial support for the implementation of the projects themselves instead typically can be found using European funding (the city of Modena was supported by ELENA[10] which is a technical assistance facility managed by the European Investment Bank (EIB) and funded by EU budget). Projects seeking funding must incorporate energy efficiency, local renewables and clean urban transport.

According to the National action plan by 2020, for which renewable energy should cover:

- 6.38 % of the energy needs for mobility and transportations;
- 28.97 % of the energy needs for electricity;
- 15.83 % of energy needs for heating and cooling.

Since more than 50 % of the energy consumed in Italy can be ascribed to cities (residential and public services), in order to reach these goals, it is necessary that the Italian government devises for cities a unique strategy and targets to be reached by 2020.

---

[8] http://www.covenantofmayors.eu/index_en.html.

[9] A deeper insight into the mechanism of EPC will be given in Chap. 7.

[10] ELENA is a European project, which provides foundings to cities that invest in energy efficiency. Modena, in Italy, is a good example of the application of this project www.aess-modena.it/progetti/elena.html.

References

Costantino D, Riva Sanseverino R (2012) Città europee e piani e progetti sostenibili: il piano "The Walkable city" in Atti delle giornate internazionali di studio, Abitare il nuovo, abitare di nuovo ai tempi della crisi, 2° edizione, Napoli, 12–13 dicembre 2012

Dominici G (2012) Smart cities nuova moda o vera opportunità? In Urbanistica Informazioni n. 243

Giffinger R, Fertner C, Kramar H et al (2008) Smart cities: ranking of European medium-sized cities, Vienna

Nylund S et al. (2010) Regional Energy Plan, Energy future of the Stockholm region; http://www.tmr.sll.se/Global/Dokument/publ/2010/2010_r_energy_future_of_the_stockholm_region_2010-2050.pdf

Profiti F S (2011) Lo stato di attuazione dell'E-government in Italia, Roma

Riva Sanseverino E, Riva Sanseverino R, Vaccaro V (2012) Atlante delle Smart cities, modelli di sviluppo sostenibili per città e territori, FrancoAngeli/Urbanistica

Riva Sanseverino E, Riva Sanseverino R, Favuzza S, Vaccaro V (2014) Near zero energy islands in the Mediterranean: supporting policies and local obstacles. Energy policy 66, Elsevier

Riva Sanseverino R (2013) Interview in Eta Beta (Italian)—Da Waze a MeTwit con le app la città diventa smart—Raitv in www.rai.tv/

Siemens (2012) Efficientcities, città-modello per lo sviluppo del Paese, www.siemens.it/cittasostenibili

Stockholm City Council (2010a) The Walkable city, Stockholm; city plan, March 2010 http://international.stockholm.se/Future-Stockholm/Stockholm-City-Plan/

Stockholm City Council (2010b) Vision 2030—a guide to the future, Stockholm http://international.stockholm.se/Future-Stockholm/

Smart Cities Atlas

2

Raffaella Riva Sanseverino

Abstract

New and fascinating examples about innovation in cities come from the real world. Every part of the industrialized world has indeed at least one case to show. But it is not just a matter of appearance, emerging economies may take the lead in the global economy growth forecast, but many studies show that the established top cities will continue to draw the wealthy for some years to come. Cities and network of cities will be the crossing point of the most important economic and financial initiatives. The catalogue of cities analysed in this section gives a synthetic representation of the measures carried out by some benchmark cities in the last ten years all over the world. Being a smart city is quite a complex goal to reach, both for cities to be designed ex novo (such as Masdar in the Arab Emirates or Caofeidian in Asia) as we will see at the end of the chapter, and for cities which have a long history behind them. Ex novo cities are also called top-down smart cities, while most of the examples reported next are bottom-up smart cities, that is, they start from existing settlements with different preconditions.

R. Riva Sanseverino (✉)
Department DARCH, University of Palermo,
Viale delle Scienze, 90128 Palermo, Italy
e-mail: raffaella.rivasanseverino@unipa.it

E. Riva Sanseverino et al. (eds.), *Smart Rules for Smart Cities*, Sxi 12,
DOI: 10.1007/978-3-319-06422-2_2,
© Springer International Publishing Switzerland 2014

2.1   Smart City Models

The urban examples analysed in this section show different ways to implement innovative approaches in cities aiming at the reduction of green house gas emissions, and more in general, at reaching an improved quality of living buildings and urban settlements for citizens.

The picture extensively given in Riva Sanseverino et al. (2012) and arising from the brief description here reported is certainly non homogeneous. The latter, especially when considering smart mobility but also the other declinations of the smartness concept, is influenced by the level of existing infrastructures, therefore by the pre-existing 'distance' between cities like (as an example) Stockholm and a Euro-Mediterranean Italian city like Palermo.[1] The deep pre-existing diversification of the urban contexts in Europe derives from different approaches that have characterized the municipal administrations in the cities. Especially in Italy, the lack of coordination in the elaboration of the urban codes together with the constraints set by daily emergencies have brought a slower development of innovative measures carried out in different ways and at a different pace in all areas of the country.

Environmental and historical heritage can indeed be a strong limitation in the integration of renewable energy sources and energy efficient measures. Their impact on the territory must be attentively evaluated in each situation and based on this some general rules for their integration can be outlined to simplify and standardize the approach. A detailed impact analysis in a Euro Mediterranean area is proposed in Costantino et al. (2012) and Vaccaro (2011).

Another general issue, underlying almost all the cited cases, concerns the economic sustainability of innovation in cities. Most experiences indeed prove that in the medium term these investments cannot create self-supporting initiatives and this is one of the added values of experimentation in cities by means of public financial support.

The public administrations have a strategic role in backing up innovation, since, as it will be shown by some virtuous examples, it can take part to the initiatives facilitating them, not taking the entire economic risk of them, but sharing it with private investors. The latter, based on their size and if involved in the management of the infrastructures may have the interest to improve the quality of the service offered and to diversify it. The public administrations should take care that private

---

[1] In Palermo it is not possible yet to talk about smart mobility: here indeed the massive use of private transportation means can be registered. The few public transportation means, obsolete and not well maintained, cannot support adequately the mobility demand of around 1 million people living the city and the surroundings. The tram system should serve the city in the future but is still on the way to be completed: only some parts are active, and these are those regained from the old railway system serving the city at the beginning of the century. The other lines that are being built should connect the city center to some neighboring areas (forecasted date of opening 2015), their completion is slowed down by the limited available resources at national level.

investors are taking actions in the citizens' interest and the presence of public institutions will increase the trust of citizens in the initiatives.

On the other hand, looking at the picture from a different and more global perspective, it is generally recognized that cities are the indisputable engines of economic growth across Europe and the entire world. In a general context of economic and technological transformations caused by globalization and by a constant process of interaction and integration (Anthopoulos and Vakali 2013), the European cities are facing a decisive challenge for the future of all European citizens: conjugating competitivity with sustainable urban development. In virtually all European countries, urban areas are the foremost producers of knowledge and innovation—the hubs of a globalising world economy. Bigger cities generally contribute more to the economy, but not all big cities do so. For cities with more than 1 million inhabitants, GDP figures are 25 % higher than in the EU as a whole and 40 % higher than their national average. The contribution of cities to GDP levels tends to level off with decreasing size. Smaller cities (up to 100,000) tend to lag behind their nations, but display average economic growth rates.

Smaller cities, where most of the European citizens live (State of European cities Ex Report Eu 2007; Efficientcities Siemens 2012; Scanu et al. 2012), are facing a hard challenge competing with larger cities, often not having availability of resources and the same organizational ability.

To apply an autonomous development starting from the resources available on the territory, and to increase their productivity, these cities have to put into evidence and value their points of strength.

"The Wealth Report (2012)", analysing the global economy and the tendencies till 2050, connects the urban smartness to the economic wealth, saying that cities that are now hardly known will be able to participate in the global economy by 2050 because are now implementing strategies to become 'intelligent communities'.

Another interesting topic of the study is the identification of 'networks of cities' as important infrastructure for the global economy.

This began in the late 1980s and now has become clear that the world's geopolitical future is not going to be determined by the combination of the United States and China. It will instead run via 20 or so strategic urban networks.

These networks have grown in importance on the back of the globalisation and urbanisation of an increasing number of economic activities. Those cities that work together begin to matter more in the global economy and in geopolitics than their respective countries. Firms that sell to other firms rather than consumers thrive on the specialised differences of global cities. Consider London, New York and Paris—they are all major financial centres, but they are specialised in very different sectors of finance. What matters to these firms is not the city as a supermarket, but as a specialised shop. By this rationale, different firms will prefer different city networks. The various city rankings and indices do measure something that matters. But for many firms, if they can avoid locating in London or New York, where costs are high, and if Copenhagen serves their purposes just as well, there is little doubt as to where they will go. The mass consumption sector is the opposite: the more cans of coke or mobile phones you can sell, the better (The Wealth Report 2012).

Being able to take part to these networks will be a key factor to lead the global economy and it will be founded on the existence of infrastructures for communication and control of urban uses and functions.

It is clear thus how this challenge can have a strong impact on any aspect of urban life quality, starting from the economy and ending up with cultural and social aspects, and of course environmental issues.

2.1.1 Northern Europe Cities

A strongly oriented and long lasting political will in designing sustainable measures for urban development is a basic element to carry out these long processes. Whatever is the political colour of the administrations, it is important to follow some general guidelines and support the projects going in the direction of sustainable urban development. In this way, such as in other virtuous examples, the city of Amsterdam,[2] the capital of The Netherlands, since 2009 has launched the Amsterdam Smart city project composed of sixteen pilot projects, in cooperation with IBM and CISCO. The project aims at testing energy-saving smart enabling technologies for sustainable choices. As a result of the starting pilot phase, the city is implementing the most effective initiatives on a large scale. At present, there are nearly 500 public charging points, and that number will grow rapidly to 1000, begin 2014.

This charging network makes Amsterdam a world leader in promoting electric mobility, while the widespread of private initiatives would allow producing and placing on the market small wind turbines and solar panels. Besides, as a result of the first phase, in Amsterdam, thousand of households are already interconnected thanks to IBM, that monitors in real time the energy consumption of private buildings connected to a smart electric grid. The aim is that to reduce the emissions as compared to the level registered in 1990, the CO_2 emissions of 40 % by 2025 and succeed, by that date, in generating 30 % of the energy required by the city from RES. To implement the entire project, articulated in technical implementations as well as in communication campaigns and symbolic actions, some local private companies also support the municipality. The Amsterdam Smart City model is indeed quite simple: at the base there are the three partners funding the initiative (Amsterdam municipality, Alliander and KPN), all showing long term economic interests (use of infrastructures) and a concurrent ambition to solve the problems of the society. Through the Amsterdam Smart City program they cooperate with other entities: Strategic partners in some specialized fields (such as Philips, Cisco, IBM, Accenture) and with small and medium enterprises.

---

[2] www.amsterdamsmartcity.com

In this way there is a differentiation between partners sharing long-term objectives (to realize the infrastructures) as well as medium term objectives (to implement strategies), and of short term such as small and medium enterprises.

The involvement of the public administration is fundamental, since it generates a positive feedback in citizens and also creates a sense of trust towards the initiative.

The first implementations can be dated back in 2009, when the public area Utrechtsestraat, through a cooperation with local entrepreneurs and citizens (Utrechtsestraat Business Association), has started to become a commercial street with a strong connotation in sustainability (Klimaatstraat, Climate Street). In the Climate street technologies, cooperative agreements and approaches are tested in order to select the most successful to make the city's (shopping) streets more sustainable on a large scale. The focus of sustainable solutions lies in three main areas: entrepreneurs, the public space and the logistics. The first two issues concerns the possibility to monitor the environment and use energy efficient solutions for lighting and heating/cooling, while the last issue concerns the waste collection using electric vehicles from a single provider, thus minimizing CO_2 emissions.

While in its Smart city strategy Amsterdam has aimed at a mix between Energy upgrading of private and public buildings (to reduce emissions also of symbolic structures as the museum of sciences Nemo) and the civic participation (as an example the cited Utrechtsestraat), cities like Gent, in Belgium, and Monterrey, in Mexico, have aimed at this latter issue to involve the citizens in the definition of inclusive strategies to improve quality of life. These cities are quite different in terms of dimensions and history but they are brought together by the continuous exchange of opinions with the citizens on ICT to improve the quality of life. Based on data collection carried out through ICT, Helsinki in Finland has improved the management policies of urban mobility and congestion management, while the Portuguese Paredes is going to install 100 million of sensors that will gather information on public lighting, energy consumption and waste disposal. All these data can dynamically influence the actions of the public administration about the cited issues.

The use of the new technologies (Batty et al. 2012) has favoured a strong improvement of the quality of life and urban mobility also in the major centre in Sweden, **Stockholm**. The Swedish capital has set out a number of initiatives in order to reach the ambitious goal to be an emission free city by 2050.

One of the strongest points in the strategy adopted by the city of Stockholm was to point on a diversified mobility system (regional metro, ethanol fuelled tram and buses), aiming at turning to zero the number of circulating private cars with an excellent public transportation system.

In cooperation with IBM, a system to detect the vehicles entering the city has been installed; the same system automatically charges the vehicles when they pass the 18 control points during the rush hours of working days.

A study has registered that after three years, the urban traffic has been reduced of 18 %, cutting down the polluting emissions of 12 %.

In the Swedish city a lot of initiatives concern the building sector. Some new projects are related to parts of the city that have been built ex novo. These projects are quite interesting from the technological point of view since they are practical implementations of sustainable and efficient buildings (the buildings at Zeroplus are energetically autonomous and also produce supplementary energy for the neighbouring areas); the Hammarby Sjöstad district in the city centre and the Royal Seaport project, that will soon be completed. The Hammarby Sjöstad, HS, project represents the first application of the "Vision 2030" plan, that has contributed to the victory of the Green Capital prize, given to the city of Stockholm in 2010.

The district has a high population density; When finished (2016) it will host 10,000 dwellings for 25,000 people and additional 5,000 working in the area over 250 hectares.

Living spaces are full of light, there are almost no private cars, a metro-tram infrastructure allows fast mobility on green boulevards, squares are pedestrianized and there are a lot of bikes.

The strategic elements of the project are those that do not appear: the waste re cycling system collects the residential rubbish through pneumatic ducts and treats them to get biogas, the latter is reused in the flats, in the remote heating central system and in the bus parking area, where also the inhabitants can buy fuel for cars.

The history of HS starts at the beginning of the nineties when Stockholm decides to be a candidate at the Olympic games in 2004. A team of architects and engineers identifies in the former industrial area of Hammarby the ideal area to build the Olympic village.

In the original project the main feature was the implementation of an eco-sustainable architecture. Stockholm does not attain the Olympic games, but the Hammarby project is not left aside, it is instead potentiated and converted to residential use.

Today the urban plan shows the main building initiative in Sweden in the last 30 years.

At Hammarby Sjöstad the waste, suitably separated, are collected in under-ground cisterns emptied by enormous extractors and sent to recycling (limiting the use of the anti-aesthetic waste bins and minimizing the waste collection).

The waste that cannot be recycled is instead carried to the local incinerator. The Högdalen co-generation plant separates combustible waste as an energy source in electricity and district heating production. The combustion generates heat enough to cover 47 % of the domestic need of heat. Another example of sustainable heat supply is the Hammarby thermal plant, which recovers waste heat from treated wastewater piped from the Henriksdal sewage treatment plant covering 34 % of the needed heat. The remaining 16 % is produced by the combustion of bio-oils. These oils originate from forest materials and grain, as well as animal fat that are not employed in the food industry. Bio-oils have almost the same properties as heating oil.

Photovoltaic panels on the roofs of the buildings instead produce the electrical energy; this RES provides the lighting for common areas, while solar panels provide half of the required heat for sanitary hot water production for residential use. Hammarby Sjöstad is provided with a closed chain re cycling system, where inhabitants 'contribute' up to 50 % to the Energy generation simply by producing waste, while the remaining 50 % is covered by RES: solar, hydro-electric and wind generation.

In summer 2005 a filling station for hydrogen cars supply was completed, there cars were already produced and circulating in Sweden at those times, such as the eco-transportation system based on ethanol buses that are replacing the entire municipal fleet.

The added value of this experience is an 'integrated' design of urban settlements namely a process involving technicians (urban planners and designers) the public administration, the citizens, the business world.

Since the very starting phases of the project, it is required to connect the master plan, the infrastructural projects and the environmental objectives. Hammarby Sjöstad, is the main expansion area of the city in recent times and it has been built following this methodology.

The different authorities, administration office representatives and private stakeholders have followed this new conceptual approach for which the first interest is the "common good".

Today the Hammarby Sjöstad project is not yet concluded, but it can be said that most of the objectives have been reached and that it can be considered a best practice showing the intrinsic potential of this design approach.

Another large realization still to be completed and at an early stage as compared to HS is the Royal Seaport again in Stockholm that will let the city reach another important target.

The area of new development will concentrate on sustainable mobility solutions, efficient building processes, and energy saving also through energy efficiency solutions.

In Stockholm 2030 will be world leader for what concerns the development and implementation of new technologies in the field of Energy and environment: since now the building of new districts, that will have the function of models at planetary scale. Stockholm Royal Seaport will be a 'shop window' for sustainable urban building where innovative and creative solutions will be experimented as well as new technologies, with the ambition that the urban district will be an environmental model at global level for the other cities. Of course the other experiences such as Hammarby Sjöstad and the other international experiences are the background for any new proposal and implementation.

The German city of **Freiburg** (Fratini 2013) has made of the environmental commitment a slogan since the sixties. It is very well known the Vaubau district—an urban area of 15 hectares—where 5,000 people live not holding a car or at least almost not using it. The shape of the district, its layout, the distribution of the functions and the public transportation means allow this.

Vaubau is an excellent example of Smart Planning, namely of an urban district organized to facilitate the mobility of its inhabitants. A tram that takes people to the city centre in Freiburg crosses the main road. The tram network covers the entire city.

The cars in the district are a few. Families, to go on vacation, use a Club Car Sharing service or share the property with other families.

Through this example we can understand how much the intelligent mobility system influence the shape and layout of the city of the future. The district is also an excellent example of smart Energy use: the buildings of the district produce more Energy than what they consume thanks to solar panels and thermal solar collectors.

The importance of the research centres as development motor in territories is proved by the experience of Tallinn, the capital of Estonia that has applied the use of innovative software to urban mobility. While Aarhus, the second largest city in Denmark is making the technological city district of Katrinebjerg a "world class environment" for research and local business. Also citizens are involved supporting their participation to decisions concerning the community. The Danish city is also building the Navitas Park, the new city hub for research and innovation that will be ended by 2014 on the renewed waterfront among low carbon emissions buildings.

2.1.2   Euro-Mediterranean Cities

Turin in Italy has candidate itself to become a "Smart City" (Pagani 2012) having a special attention to environmental issues developing low carbon technologies for uses and functions of cities. It is certainly the first Italian city that can praise itself for having taken some concrete steps towards the transformation into an eco-sustainable city, namely able to meet the inhabitants needs reducing its own carbon footprint. With the adhesion to the Covenant of Mayors, committing cities to support the European Union target to reduce of 20 % the CO_2 emissions by 2020, the city of Turin has completed the elaboration and started implementation of its own Action Plan for Sustainable Energy.

Specifically, the CO_2 emissions in the city of Turin has been reduced from 6.270.591 ton in 1991 (taken as reference year) to 5.100.346 ton in 2005, with a reduction of 18.7 % in the considered period.

The pro-capita emissions in 2005 have reached 5.6 ton/inhabitant. The city mainly works on three different directions: the Energy conversion, the remote heating and that of sustainable mobility.

In Italy, Parma has signed an agreement with IBM for the realization of video-branches installed along the city roads where citizens can remotely conclude administrative records. Improving the relation between citizens and public administrations in Italian cities, where the level of bureaucracy is quite high, is one of the main targets of 'smartness' oriented actions.

Among the Italian cities also Palermo, in Sicily, is defined a city in transformation in the already cited study about *Cities and infrastructures for growth* carried out by Cittalia-Anci research and Siemens Italia (Siemens 2012).

Palermo has recently undersigned the Covenant of Mayors and has presented the Action Plan for Sustainable Energy (Piano di Azione per l'Energia Sostenibile 2013), where measures to reduce emissions and improve urban mobility are designed (Riva Sanseverino et al. 2013).

The city moreover is partner with the University of Palermo and some SMEs (Italtel, Muovosviluppo, CNR-ITAE Messina) in a project (i-Next) financed by the Ministry for Research and Education in the call 'Smart Cities and Communities' that should support different measures in many fields (energy efficiency, electrical smart grid, etc.).

Malaga, is a leading city in the project CAT-MED[3] (Changing Mediterranean Metropolises Around Time), launched for the first time from the same Malaga in may 2009, with the aim of identifying operational solutions that can influence concretely on the habits of citizens so as to limit the environmental impact of urbanization and the green house gas emissions.

Eleven Mediterranean cities have therefore decided to concentrate their projects and efforts to prevent natural disasters through the promotion of a sustainable multi-functional urban model. The project CAT-MED gathers the cities of Athens and Thessaloniki (Greece), Barcellona, Malaga, Valencia and Seville (Spain), Rome, Genoa and Turin (Italy), the community Pays d'Aix and Marseille (France). The Institute for the Mediterranean technically supports them.

The objective of primary importance is that to show, through a trans-national experimentation, the importance and the strategic value of actions implemented in a coordinated manner between cities in order to prevent natural threats deriving from global warming.

The implications of this ambitious goal are that at city level, each city must promote a sustainable urban model influencing the behaviours of citizens and in the administrations.

This is the first experimental step to create a strategic convergence between cities that, in the long term, will be able to face the environmental challenges in the Mediterranean Sea in a more efficient way. In the CAT-MED vision the convergence process must be accompanied by the measurement of the objectives through suitable indicators. The indicators involve the 5 pillars of sustainable development: environmental protection, social cohesion, economic efficiency, territorial approach and governance processes. Grouped by a centralized GIS (Geographic Information System), the comparative analysis system allows the attainment of goals in fighting against climate changes and lets people know the relative position of each city taking part to the project as compared to the optimal values of the indicators.

---

[3] www.catmed.eu

For each indicator, target levels are thus defined, and to these levels cities will have to refer in fighting against climate changes. Another further objective is the definition of initiatives finalized at making concrete such 'convergence'. To reach this target experimentations in all cities involved in the project have been developed in a coordinated manner.

Each city indeed is developing a pilot project for the definition of a 'sustainable district', involving groups of the main stakeholders. Such groups are connected at trans-national level to exchange operational solutions identified in each pilot area. The considered path will create a common model implementing principles and the identified measures, so as to edit methodological guidelines to sustainable districts. The guide will provide solutions to a large number of problems concerning the sustainable urban development, so as to prevent the risks deriving from climate changes.

2.1.3 Cities of the World

In the US the alliance with the big innovation players has brought positive consequences for the smart growth of many urban settlements: in Seattle, Washington State, the partnership with the fellow citizen Microsoft has allowed the inhabitants to trace on-line their own Energy residential Energy consumptions (Nextville Editorial Staff 2013), contributing in this way to the attainment of energy saving targets set by the Climate Action Plan while in the close city of Portland the partnership with IBM has allowed to analyse the data of different urban phenomena to evaluate the possible interconnections and integrated actions for improving the urban environment and the quality of living.

Among the Asian cities[4] probably Singapore[5] is the smartest especially for what concerns the transportation system with a modern metro (System Mass Rapid Transit) in continuous development, connecting all parts of the city and the surrounding areas in all directions. The opening of the tunnel of the Singapore's Circle Line, a fully automatically operated metro line is a recent event of a never ending work of reinforcement. The public transportation system in Singapore is efficient and well organized and allows reaching any point of the city.

Since not so long has started its operation the new line connecting the city to the airport of Changi. In each station S.M.R.T. it is possible to buy a one-way ticket at the automatic distributors or using the Ez-link card, easily rechargeable at all the metro stations. It is a large expense to buy and drive a car in Singapore because the government with many measures controls the number of vehicles circulating in the state, in order to limit the air pollution and avoid traffic congestion. The large import taxes, the registration taxes and the payments for licenses put a severe constraint that limits as a matter of fact the number of citizens that in Singapore can own a car. The toll system (congestion pricing) called Electronic Road Pricing

---

[4] www.futuregov.asia/

[5] www.senseable.mit.edu/livesingapore/; www.ida.gov.sg

(ERP) in 1998 launched the adoption of new technologies with the commitment, for each car entering Singapore, to own a system allowing the identification at one of the 69 checkpoints and entries of the city.

The ERP is an electronic toll system and to enter the city and in the Restricted Zone, with differentiated fares according to the type of vehicle, the times of the day, the area and the type of road.

The boundaries of these toll areas can be easily identified through blue structures on top of the roads. Lighting signals indicate when the ERP is active. Large posters close to the structures indicate the fares. In each vehicle, on the anterior part, there is an electronic system called In-Vehicle-Unit or IU. When the vehicle enters the Restricted Zone, a Cash Card, already inserted in the electronic system will signal the fare that will be automatically charged to the card.

The Massachusetts Institute of Technology (MIT) and the National Research Foundation of Singapore have announced a joint project for the development of new models and tools for planning and operation of the future urban transportation system, such initiative started in 2010. The project is centred on the development of the SimMobility, a simulation platform o fan integrated model of human activities, business, use of the soil, transportation, environmental impact and Energy consumption.

The project researchers are planning to use this platform to design and evaluate new mobility solutions in urban contexts in the city of Singapore.

A project of the joint venture Renault-Nissan is that to take the largest role in the world production of zero emission electric vehicles. The Alliance has formed a partnership with the Energy Market Authority (EMA), the Land Transport Authority (LTA) and the Economic Development Board (EDB) of Singapore to explore the development of Zero Emission Vehicles Program.

The city of Singapore is besides quite adequate for Electric Vehicles mobility due to the limited size of each vehicle, the urban context supplied with a widespread electric power network and the existing technological infrastructures for data transfer. The strategic importance of Asian cities, the land's geography and its commitment in environmental subjects allow a large development of the electric vehicles market on a large scale.

The experiences related to the cities built ex novo through a top down approach are also very interesting since they are a real world test system of the most varied technological solutions with some excess typical of frontier research applications.

These experiences go from Masdar[6] city, see Fig. 2.1, to other entirely new districts (Caofedian New District). Masdar city (from Arab: Madīnat Madar literally means source city) is a city planned at Abu Dhabi in the Arab Emirates. Projected by the English designer Norman Foster the city will count exclusively on solar energy, with a zero emissions economy and a zero waste society. The company Abu Dhabi Future Energy Company (ADFEC) led by the sheik Mohammad Bin Zayed Al Nahyan manages the initiative.

---

[6] www.masdar.ae/

Fig. 2.1 The pillars of sustainable city according to the Masdar city model

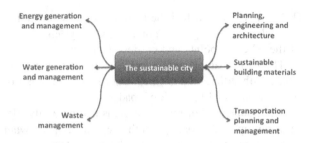

The city presents itself on the Internet as "A place where businesses can thrive and innovation can flourish" and "it is a model for sustainable urban development regionally and globally, seeking to be a commercially viable development that delivers the highest quality living and working environment with the lowest possible ecological footprint".

The project started in 2006, while the building of the first constructions started in 2008 and is still to be ended.

Madinat Masdar—a giant project of 22 billion dollars for the energy company Masdar—is a city where the common cars will not be able to circulate, and will be replaced with around 2.500 zero emissions shuttle buses that will travel over 150.000 different routes every day. The energy required to supply this 'eco-toy' comes from wind, solar and solar thermal systems, that will allow a cost reduction in the next 25 years of 2 billion dollars of oil.

The city will host, at the beginning, 50.000 people, 1.500 enterprises and, over all, the futuristic Masdar Institute of Science and Technology, high education pole realized in cooperation with the Massachusetts Institute of Technology and dedicated entirely to study and research in the field of renewable energies. Masdar City aims at becoming one of the most sustainable urban settlements in the world.

The city will host International companies in the field of renewables and of sustainability products. General Electric with its centre Ecomagination for the development and widespread of new and innovative technologies is a strategic partner in Masdar City. The project involves other partners such as BP, Royal Dutch Shell, Mitsubishi, Rolls Royce, Total S.A., Mitsui, Fiat, and the German Conergy, that is planning a solar thermodynamic power generation plant of 40 MW rated size.

References

Anthopoulos LG, Vakali A (2013) Urban planning and smart cities: interrelations and reciprocities, Springer, Berlino

Batty M, Axhausen KW, Giannotti F, Pozdnoukhov A, Bazzani A, Wachowicz M, Ouzounis G, Portugali Y (2012) Smart cities of the future in European physical journal special topics 214, pp 481–518. in Springerlink.com

Costantino D, Ippolito M, Riva Sanseverino R, Riva Sanseverino E, Vaccaro V (2012) Sustainable integration of Renewable Energy System in a Mediterranean Island: a case study in Sustainable Energy and Buildings, Springer, Berlino

Efficientcities, città-modello per lo sviluppo del Paese (2012) Siemens in www.siemens.it/cittasostenibili

Fratini F (2013) I quartieri sostenibili di Friburgo in urbanistica informazioni n. 248

Nextville Editorial Staff, Vademecum Nextville 2013 (2013) Efficienza energetica, gli incentivi per il risparmio energetico, le rinnovabili, termiche e la cogenerazione, Milano

Pagani R (2012) L'urbe diventa smart, in Quale energia? Aprile-maggio 2012

Piano d'azione per l'Energia sostenibile (2013) Palermo

Riva Sanseverino E, Riva Sanseverino R, Vaccaro V (2012) Atlante delle smart cities: modelli di sviluppo sostenibile per città e territori, Franco Angeli/Urbanistica

Riva Sanseverino R, Riva Sanseverino E, Costantino D, Vaccaro V (2013) The actionplan for sustainable energy in Palermo: action and measures for a city in transformation in XXVIII INU Congresso Nazionale Istituto di Urbanistica, Salerno 24–26 Ottobre 2013

Scanu L, Troncarelli D, Venturini L, Frau A (2012) Decentramento energetico e smart city: nuove opportunità di crescita per le utility italiane, in Harvard Business Review, supplement a—special practice accenture, pp 56–63

State of European Cities Executive Report, EU (2007)

The Wealth Report (2012)

Vaccaro V (2011) Nuovi modelli per la città contemporanea: la smart city—Un progetto per l'isola di Pantelleria, Tesi di laurea, relatori: D. Costantino, R. Riva Sanseverino, correlatore: E. Riva Sanseverino, Università degli studi di Palermo

Vianello M (2013) Smart cities: gestire la complessità urbana nell'era di internet, Maggioli Editore Pionero

The Integration and Sharing of Resources for a New Quality of Living

3

Raffaella Riva Sanseverino and Salvatore Orlando

Abstract

In this chapter, the issues of sharing of information and of information technologies use are dealt with from the juridical point of view, through a discussion about some general problems characterizing the relevant juridical debate. Then the urban forms and functions of the smart city are presented. Information technology can interact with the operational problems of the city and the use of environmental resources (energy, soil, water) is the leading parameter with which the urban and building transformations must be carried out. In this chapter, the complex issue of how to share the urban spaces and functions and to what extent such sharing influences the energy consumption is dealt with.

R. Riva Sanseverino (✉)
Department DARCH, University of Palermo,
Viale delle Scienze, 90128 Palermo, Italy
e-mail: raffaella.rivasanseverino@unipa.it

S. Orlando
Department of Law, Sapienza University of Rome,
Via del Castro Laurenziano 9, 00161 Rome, Italy

S. Orlando
Law Firm Macchi di Cellere Gangemi, Via G. Cuboni 12, 00197 Rome, Italy

E. Riva Sanseverino et al. (eds.), *Smart Rules for Smart Cities*, Sxi 12,
DOI: 10.1007/978-3-319-06422-2_3,
© Springer International Publishing Switzerland 2014

3.1   What Legal Framework for Smart Cities?

According to some surveys, smart cities spending will reach globally US $20 billion by 2020 (Clancy 2013) Still, an overview of the law-systems of the many jurisdictions where smart city projects and investments are in place reveals that there are no such things as "smart-city statutes", i.e. laws governing smart cities as a single phenomenon in an organic and comprehensive way. One may ask whether similar statutes are necessary, and the question is, effectively, of a preliminary importance. More precisely, it appears relevant to discuss in the first instance whether reasons exist for justifying the intervention of legislators in this sector, similarly to what one expects to discuss in relation to any other business sector. A connected question is whether smart-city statutes are possible. This question requires assessing to what extent legislators may efficiently intervene for governing smart cities, in case reasons are found to justify their intervention[1].

Based on the reasons that I will try to summarize below, it appears sensible to conclude that smart-city statutes (i.e. comprehensive statutes governing all aspects of smart cities) are neither necessary nor possible, while sound reasons seem to exist for justifying the intervention of legislators or governmental authorities in certain fields of law and primarily in connection with the new challenges in security and privacy raised by smart cities.

3.1.1   Definition of Smart Cities and Central Role of ICTs: Examples

The starting point is the definition of smart cities. What do we exactly mean when talking about "smart cities"? To the extent necessary for a legal analysis, I will use a broad definition and an operational notion of smart cities. A broad definition of smart cities has been given as follows: "We believe a city to be smart when investments in human and social capital and traditional (transport) and modern (ICT) communication infrastructure fuel sustainable economic growth and a high quality of life, with a wise management of natural resources, through participatory governance" (Caragliu et al. 2009). This definition is based on an operational notion, which, relatively to medium-sized European smart cities, has been proposed using six characteristics: Smart Economy, Smart People, Smart Governance, Smart Mobility, Smart Environment and Smart Living (Giffinger et al. 2007). These characteristics are, in turn, the hives of a number of factors, described in the form of a Fig. 3.1 as follows.

We get from the above definitions that smart cities are actually denoted by a political vision, and more precisely by a set of aims and objectives, which need to be factored into coherent programs of urban and service improvements. If we continue our analysis (Cassa depositi e prestiti 2013), we also realize that information and communication technologies (ICTs) are necessary for smart cities, because the

---

[1] Salvatore Orlando is the exclusive author of Sect. 3.1.

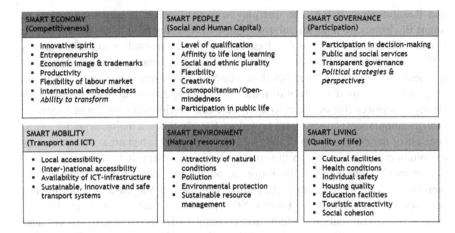

SMART ECONOMY (Competitiveness)	SMART PEOPLE (Social and Human Capital)	SMART GOVERNANCE (Participation)
• Innovative spirit • Entrepreneurship • Economic image & trademarks • Productivity • Flexibility of labour market • International embeddedness • *Ability to transform*	• Level of qualification • Affinity to life long learning • Social and ethnic plurality • Flexibility • Creativity • Cosmopolitanism/Open-mindedness • Participation in public life	• Participation in decision-making • Public and social services • Transparent governance • *Political strategies & perspectives*

SMART MOBILITY (Transport and ICT)	SMART ENVIRONMENT (Natural resources)	SMART LIVING (Quality of life)
• Local accessibility • (Inter-)national accessibility • Availability of ICT-infrastructure • Sustainable, innovative and safe transport systems	• Attractivity of natural conditions • Pollution • Environmental protection • Sustainable resource management	• Cultural facilities • Health conditions • Individual safety • Housing quality • Education facilities • Touristic attractivity • Social cohesion

Fig. 3.1 Characteristics and factors of a smart city (reproduced from Giffinger et al. 2007)

programs realizing the aims and objectives, which are characteristic of smart cities, need to make use of ICTs. The above indicates at the same time that ICTs are not sufficient to define a smart city, but still are essential to smart cities. In other words, if it is true that smart cities are not just "more investment in ICT", it is also undoubtedly true that smart cities do necessarily involve more investment in ICT.

This latter observation applies to practically all domains characterizing smart cities.

If we consider, for example, the Smart Living factors (as listed in Fig. 3.1), we realize that the ability to improve Quality of Life is dependent on the ability to efficiently combine technology and information (data) to reach the stated objectives of improvement. For instance, the objective to improve quality of life in connection with the Health factor (which comprises inclusion and assisted living) is dependent on the ability to provide healthcare and other services using innovative ICTs, since information and communication technologies are a fundamental means for helping the containment of healthcare delivery costs while maintaining or increasing levels of quality of care and safety. Data transmission technologies enabling remote diagnosis, including in wireless and mobile scenarios, are and will be more and more important in the future for developing in-home monitoring, as well as for offering more independence, security and autonomy to senior people and people with disabilities.

Likewise, most of the objectives attached to Smart Mobility (and notably the general objective of optimizing trip planning and management) appear to be dependent on ICTs. For instance, public parking spaces can be more efficiently managed, by guiding drivers to nearby free parking places through portable or car-mounted devices providing accurate location information. It has been estimated that by lowering of only 3 min the average time needed to find a public parking place, the associated reduction in terms of CO_2 emissions in a town like Barcelona would be of 400 tons/day (Expert Working Group on Smart Cities Applications and Requirements 2011). The interplay with the general objectives attached to Smart Environment, and particularly to those of reducing pollution, is also evident. Urban traffic control systems developed and adapted into intelligent traffic systems (ITS) capable of tracking cars location in real time, and adapting traffic management to current and predicted conditions, are instrumental too. They are envisaged to be used not only to reduce traffic congestion but also to serve efficiency of services (for instance, to set up fast lane corridors for emergency services such as ambulances, police or fire brigades). Clearly this is a major target for cities, because of the concentration of the global population in urban areas. To confine us to comment on the European dimension of this statement, it is sufficient to note that approximately 80 % of the European population lives in urban areas. Complementarily, dynamic carpooling systems (Correia and Viegas 2009), or the ones developed in the WiSafeCar Project 2013, provide a means to optimize the utilization of transportation systems for commuters living in nearby places and sharing a common destination. In brief, platforms allowing to share information (location, weather and traffic data) among transportation system operators, urban districts and passengers are needed in smart cities.

As to Smart Environment, and to energy consumption particularly, the combination of smart processes (e.g., demand side/response management and real-time consumption management) and smart technologies (e.g., smart meters and intelligent home energy management devices) is considered as an opportunity to enable energy efficiency and savings to be achieved in the residential and business market. In this context, intelligent systems and integrated communication infrastructures are required to assist in the management of the electricity distribution grids. Smart grids are actually seen as a major opportunity to merge power and ICT industries and technologies to satisfy almost all stated targets, and ICTs are, again, of essence, because of the need for the underlying communication, consisting in sharing information among consumers, producers, and the grid.

3.1.2 Cities' Transformation from Service Providers to Platform Providers

The above observations indicate a process of transformation (AGCOM 2012; Evans 2011) relevant to the role of cities as service providers. Cities are increasingly moving from being service providers to platform ones, by setting up and making available to users at large infrastructures enabling the development of a broad range of public and private applications and services.

Two aspects of this process appear worth factoring in our analysis on the legal framework for smart cities. The first one is the increasing importance of information and communication in the process that will grant cities a new role as platform providers. The second aspect, which is connected to the first one, is the increasing importance of private law (compared to administrative law) for cities, as they start realizing the many legal issues relevant to the processing of the information made available in the platforms. This second aspect, in turn, will progressively make local governments and administrative bodies confront with a new legal environment, which does not allow them to use authoritative and discretionary powers to the extent admitted in traditional administrative law. This is particularly true with respect to personal data, the regime of which is mandatorily provided by law provisions laying on principles posed by national and international (EC 2011; EC 2012) sources (such as, in the European Union, the EU directives and regulations dealing with data protection). Other sectors of private law at large that are increasingly relevant to smart cities in connection with the various communication issues involved by smart cities are industrial and intellectual property law (including copyright law) and commercial law (including the realm of e-commerce).

Certainly there are ample fields relevant to smart cities programs where administrative law needs to apply and develop, especially in connection with urban programs, and also a need for public incentives (through financial and investment plans), however, the essence of smart cities as a matter of regulation in every jurisdiction, its "raw material" if we can so say, seems to be information, which calls for an assessment of the nature and limits of the intervention of legislators, especially in connection with data protection and data security.

On the contrary, because of the wide range of applications and procedures that are commonly referred to when talking about smart cities, it is hard to imagine single statutes capable of governing all legal aspects relevant to smart cities.

3.1.3 The Need and Rationale for a Legal Intervention in the Data Protection Sector

The above observations may induce to believe that the issue of data protection in connection with smart cities is essentially a compliance issue, consisting of assessing on a case-by-case basis compliance of smart-cities programs with mandatory provisions. This approach would be however misleading under many respects.

First of all, such approach could not but be based on the assumption that law provisions exist efficiently governing the specific privacy and data security issues relevant to smart cities, while, this is not exactly the case. It seems, on the contrary, that smart cities deserve some special consideration by legislators (Orlando 2012) as far as the legal regime of data protection is concerned, in order to capture and govern, through new law provisions, the specific challenges raised by smart cities to privacy and security.

Law provisions (Mattei 2011) dealing with data protection, far from being stable and immutable, are subject to continuous changes and adaptations. Generally speaking, the main drivers of changes to law provisions dealing with data protection are of two kinds: (i) technological changes (to make simple and self-evident examples, it is easy to understand the peculiar problems in data security and data protection historically triggered by the advent of digital technologies, and more recently by cloud computing and the internet of the things), and (ii) the consideration of stated interests or principles that, in certain areas of application, may be deemed worth of protection and, if conflicting with those underlying the data protection legal regime, suggest to provide for differentiated legal treatments, i.e. special provisions.

This latter phenomenon is observable for smart cities because of the potential for data protection to put at risk the objectives of smart cities, and because of the importance of the objectives of smart cities (such as the objectives of reduction of pollution, efficient health care services, efficient consumption of energy, reduction of traffic congestion etc.) which make them appear in principle worth of protection. Because of their nature as objectives of a "general" or "public" interest, the smart cities objectives appear therefore to deserve a special consideration by legislators. In smart cities, in other words, data and personal data are processed not only or not exclusively to the advantage of companies (i.e. to pursue a private interest) but also for the achievement of the above mentioned objectives of a general or public interest. This makes the data processing in smart cities peculiar and worth of special consideration.

For the same reason, to apply to smart cities all of the existing law provisions whereby the data processors (i.e. those subjects that process third party personal data) are seen as subjects exclusively pursuing interests of a private nature (business interests) without a direct impact on the achievement of objectives of a general interest, seems not necessarily, or not entirely, appropriate.

It appears therefore that legislators have some work to do in this direction, and that special legal provisions need to be enacted to achieve satisfactory compromises between the conflicting interests involved by smart cities, i.e. the individual interests of the data subjects (traditionally protected to the maximum extent by data protection legislation), on one side, and the public or general interests envisaged by smart cities' programs, on the other side.

Another interesting issue to discuss is the rationale for advocating data protection and data security in connection with smart cities.

While it is certainly true that inviolable personal rights shall form the theoretical basis for such protection (and also a firm one), some "commercial" or practical reasons can also be identified for advocating a certain degree of data protection and data security in connection with smart cities. This second type of reasons revolves around the appeal of smart cities programs for users, and, ultimately, their chances of success.

Generally speaking, it is a common observation that users implicitly expect systems to be secure and privacy preserving. Systems that are perceived by users as insecure or threatening users' privacy have poor chances of success, and, in any case, minor chances to establish successfully in the market compared to those systems that are able to gain users' consent and trust. This general observation seems to indicate that acceptance by users of security and privacy-preserving procedures is a target for smart cities, and that research needs to be developed in this direction too.

According to an authoritative report (Expert Working Group on Smart Cities Applications and Requirements 2011), research challenges in this sector can be classified into the following aspects: handling of the increasing complexity of distributed systems from the security perspective; identity and privacy management, where, e.g., pseudonymisation must be applied throughout the whole system, in order to separate the data collected about a user from the user's real identity; integration into systems of security technologies, e.g., advanced encryption and access control, and intelligent data aggregation techniques.

In conclusion, advocating data protection and data security in connection with smart cities is not only a due exercise in view of the protection of individuals' inviolable rights but also an exercise for increasing the chances to establish successfully smart cities' programs in the market, and, in the end, a competitive factor. At the same time, it appears that, because of the characteristic typical of the data processing relevant to smart cities programs, consisting in the aim at achieving results of a general or public interest, the regulation and government of data protection and data security in this field deserve special consideration by legislators and governmental authorities.

3.2   Sharing and Integration in the Smart City

In the intelligent city, it is desired a strong integration among urban functions, both traditional as well as innovative (energy, mobility, etc....), that increasingly play an important role. The information technology is an exceptional implementation propeller able to connect and tie up the urban innovations that can develop in the contemporary scenario (Cittalia—Fondazione ANCI ricerche 2012). The sustainable use of environmental resources and the widespread and integrated use of renewable energy appear to be a necessary condition in the city of the future. Besides the process is very broad and involving different areas of the city and mostly the people (smart people), end users of urban transformations. Some functional areas, where it is concentrated in a timely innovation in smart city are mainly: the transport sector (smart mobility) in the first place and the energy sector (smart energy)[2].

In this context of great changes on issues that affect the daily lives of all of us, citizens must do their part, being the main actors in a new process. They are a key component of smart city: their choices, their behaviours and their education are essential to bet on the future innovation processes of the city. But this step is not obvious, because it is often very difficult to change people's habits (especially if they are already adults). The "smartness" here must be understood as a skill or attitude of a community, in a time of scarce resources, to share and integrate, in a competitive and strategic way, functions and services, giving rise in many cases in habits and established procedures changes. The process of listening and participation of the citizenship is the basis of policies in some Italian municipalities to improve the quality of life of citizens, as well as the quality of the urban fabric through a better organization of schedules of services in the area (Time Plan and

---

[2] Raffaella Riva Sanseverino is the exclusive author of Sect. 3.2.

Schedules, City of Bolzano, 2005).[3] The sharing of projects in the smart city is practiced by different means: the smart city often uses Urban Centre that serve as real bodies to disclose and make known to the community, projects and new initiatives. Sharing in some areas necessarily happens through technology (energy management systems or traffic control through digital systems already in place in many European cities). The following paragraphs deal with the theme of sharing services in the city of the future. Finally will be addressed another important aspect of the smart city: the integration, which leads many innovative processes and that, lends itself to different interpretations and applications.

3.2.1 Sharing in the Digital Age

In line with the trend that stems from social networking sites—Facebook and Twitter in the first place—sharing is a buzzword of the contemporary era: today in a more immediate way than in the past you share an experience, a book, a sentence, a journey but also a pair of shoes and a dress, basically everything (Lamborghini and Donadei 2006). The teenagers before buying a shirt or a personal item, after taking a picture of the item to be purchased with the iphone, do a survey among her friends on facebook to see how many "likes" records (and only after the purchase).

In real time we are able to communicate with those we want: in daily practice we can record information and exchange data and images with people who are physically far away.

This is the era of Internet, web and shared communication: images, phrases, ideas that sail into the wild. The arena of digital sharing remains a world of open networks and increasing exchanges, new knowledge and a variety of content through a grid of nodes, consisting of millions of PCs scattered throughout the world. The transformation of the scenario results in new structures, which are difficult to control, but the real victory lies in playing the game all the way betting on innovation and creativity.

In this way sharing is connected to sustainability and becomes a practical driving force in the projects of smart cities.

Sharing can infact become the tool that allows achieving certain objectives (bringing sustainability, but also economic benefit as we shall see) the sharing of projects, involving citizens, who are the primary users of urban transformation.

In this way, the municipal government, before starting any project, try to explain looking for the involvement of citizens and of all the local players. This is the new dimension of *urban governance*, a term in use since the seventies and eighties in emerging countries, is the new watchword. Governance policies are based on the sharing between multiple actors of the decision-making process for the organization and development of the territory. Participation must be for this reason extended and enlarged to everyone.

---

[3] http://www.comune.bolzano.it/index_it.html

The theory is of little importance, the practice and the *good practice* are here more important: the positive signals coming from on going field experimentations. In terms of waste treatment, many cities (including Italian cities) went on working on the theme of recycling, initiating educational projects right from the childhood-schools to make children understand how important recycling materials is, especially for the environment. Creative workshops with recycled objects, exhibitions in schools: all this to make sure that behaviours are guided since the child-hood and may arise following a "bottom" and the gradual approach.

Citizen participation is essential in the experiences that relate to the town: in some cities, the separate collection of waste becomes a way to lower taxes for citizens computed is according to the amount of separate waste produced (kg)/household. Only in this way the process becomes shared and democratic, citizens become part of an overall process that involves them directly.

Slowly, people's attitudes change; so that today in many cities it is possible to buy detergents and milk directly "on tap" without wasting new containers, in order to have a direct savings on products purchased without producing new unnecessary plastic containers that should then be disposed.

3.2.2   Sharing and Integration of Urban Functions: Mobility and Energy

The share of urban functions such as transport systems, as will be illustrated in the case of Freiburg, is a step forward towards sustainable dimension of some (mobility) urban functions.

The car has always been considered a good that is linked, in the collective imagination, to private property, a personal item.

In the modern era, cars have been chosen with nice and fashionable colours as well as all its features: model, size, brand, etc. elements that have made it progressively feel only our own.

In Italy, as an example, during the economic boom of the post-war period has been linked to the spread of utilitarianism—the myth of the Fiat Cinquecento—which marked a significant period of the Italian history.

What will be outlined in this section concerns the example of Freiburg: a new approach to the transport system of the city. The car is no longer a private good, as it was said above, it belongs to the community and it is shared between citizens. The ultimate goal of the community is the environmental advantage that thanks to this practice can be derived.

Sharing in fact becomes a practice of environmental saving in Freiburg, experiencing since some time a shared environmental project, car-pooling or car sharing. The city of Freiburg is a small university town, considered the ecological capital of Germany under the policies she has pursued for more than thirty years with a high level of citizen participation in the definition of the strategic choices of urban governance.

The choices made by the administration, particularly as regards urban mobility (pedestrianization, low time parking areas for cars with expensive rates, car-free neighbourhoods, integrated ticketing system for public transport), made it possible to maintain the use of private cars to the levels of the early seventies. Today in Freiburg 70 % of urban trips are made on foot, by bicycle or by public transport and 90 % of the city's residents live in areas with limited traffic.

In some districts, there are very few cars that circulate and public transport, very efficient, they are definitely the fastest way to reach any place of the city. Bicycles are another means used by the inhabitants: bike paths in the city parks and in the city centre become an alternative way to cross the city and reach different places. The low impact integrated transport system is completed with car-sharing or rental car services.

If Freiburg resides one of the most innovative companies in the field of urban mobility and renewable energies, this is due not only to an undisputed technological advancement, but especially to the forward-looking participatory policy actions.

The Vauban district in Freiburg is spread over an area of 30 ha and once hosted military barracks. It was created during the Nazi period and subsequently used by the French army since 1945 until the fall of the Berlin Wall. In 1992, the Vauban barracks were returned to the city, which has carried out an ambitious plan for environmental restoration.

In about 10 years 2000 housing units to accommodate the current 5500 residents have been built. All the residents have sold their cars after they moved into the district. Freiburg is a cyclable city since several years: 28 % of journeys are made by bicycle for a total of about 211,000 trips every day. A very high average, considering the total number of inhabitants and assuming that on average every citizen performs a cycling tour at least once a day.

But in Vauban citizens and administration have gone above and beyond. In the streets are allowed to park only private cars for loading and unloading of goods, while the car park is quite expensive. Parking along the only avenue where it is allowed, the Vaubanallee, is quite expensive. Same thing goes for the families of the residents who are allowed to purchase a single parking space. Some associations allow, at lower costs, to build a garage in the property of the association. But many people now do not feel the need of owning private vehicles.

Currently 70 % of households do not own a car (on average Italians for example own 1.5 cars/household). It is interesting to analyse the relationship people/cars in European cities and in the Italian cities: Paris and Amsterdam is 25 out of 100 cars, 70/100 in Rome, 62/100 in Turin. In Vauban some families own a car jointly, while for those do not, there are efficient car-sharing services. In this context, perhaps the most innovative aspect is the active involvement of citizens in the transformation of the district.

The adoption of strong ecological criteria for mobility and building, as well as the attention to the most vulnerable social groups were the basic criteria of the founding group of citizens of the Vaubau district.

In the autumn of 1994 a group of motivated citizens with very few resources created an association—Forum Vauban—to promote the ideas of citizenship on the future to be developed over a former military area. The subsequent year, in 1995, the City Council recognized the Forum Vauban, as a subject partner to the table for the participatory design of the new district. The project development and implementation lasted year sand involved individuals, committees, companies, political parties and resulted in a huge variety of initiatives and concrete achievements.

Returning to the main analysis, an idea or a principle can be actively shared through the Fair Trade Groups[4] (FTO) (Saroldi 2001; Montagnini and Reggiani 2010; Rossi and Brunori 2011), increasingly popular organizations of consumers. Such organizations are now extended to other spheres business, as will be explained with regard to smart energy, buying groups *tout-court*, "that may not have ethical implications, but are just a tool, but it is also important the reference to the importance of social and human relations or the connection with the environment or with the traditions of farming and food".

The criteria that guide the selection of suppliers (although different from group to group) generally are (Saroldi 2008): product quality, dignity of labour, respect for the environment. In general, the groups (Valera 2005) also pose great attention to local products, organic foods or equivalent and packing. Secondarily, but equally fundamental, is the reference to the importance of social and human relations or the connection with the environment or with the farming and food traditions.

In the constitutional document of the FTO, the basic principles are environmentally friendship, solidarity, equity and sharing. The structure of the FTO is highly flexible and articulated.

In the wide panorama of FTO there are very different associations leading to different interpretations. Organising procurement and internal communications is equally variable, for example, related to the number or type of participants, place, or the choices of the Organisation. Often purchasing goods FTO use software created specifically to manage the collective orders (FTO management software).

The purchasing FTO in general is a set of consumers who purchase a particular type of goods (usually food, but not only) directly from the manufacturer without going through intermediation, such as shops or wholesalers that drive up the final price of the product.

The ADDRESS[5] Project (Active Distribution network with full integration of Demand and distributed energy RESourceS) outlined in Petroni (2011) is a large scale European project in Framework Program 7 in the area "Energy" for the development of active electrical energy distribution grids. The project started in 2008 and lasted 4 years, ending in may 2013.

Coordinated by ENEL,[6] the Italian electricity distributor, the ADDRESS Project was carried out by 25 partners of 11 European countries: universities, research

---

[4] Constitutional document FTO in Italy—http://www.retegas.org/. In Europe the association of some Fair Trade Groups can be found at www.urgenci.net.

[5] www.addressfp7.org

[6] www.enel.com

centres and European companies, mostly distributors and sales company of electrical energy as well as producers of domestic appliances and electric components.

The research project has developed solutions to let small and medium consumers to actively take part to the energy market, offering modulation services for their consumptions and selling the energy produced (pro-sumers).

The ADDRESS project aims to study the active participation of consumers in the new scenario and intelligent networks. This theme is gaining more and more interest and, in particular, in the last year, saw the birth of new initiatives. The issue faced by ADDRESS are synergistic with the broader theme of intelligent networks (Smart Grids).

There are several dimensions addressed in the project:

- operation of the network in the new scenario, with the involvement of consumers, the aggregator and distribution companies;
- socio-economic aspects in order to understand the needs of consumers and look for the most appropriate solutions;
- telecommunications to study and define an architecture of communication that makes it possible the real-time interaction between customers and other market players;
- measurement and management of smart appliances, storage and generation for domestic purposes;

In 2011, the first tests were performed in the laboratory prototypes, followed by the real demonstrations that have been conducted with the involvement of consumers in some European countries (Spain, Italy and France).

ADDRESS deals with residential or small commercial exercises, consumers connected to the low voltage network. The project concerns the habits of European citizens and has the objective of establishing the application and active participation in the system by studying how they can change the energy consumption and user flexibility.

The consumption of this type of customers can be made more flexible from the point of view of the time, of the amount (consuming less or at different times) controlling and managing the domestic appliances (for example, the washing machine, dishwasher, air conditioning, the electric heating).

This project team actually shows the composition of the electricity market in Europe dominated by the large electricity distribution companies; the idea conveyed through the project is that the service deployed by the new actor, the aggregator, is still a distributor owned function. The aggregator indeed draws up cheap packages and offers to consumers, playing an intermediation role, between the consumers and the distributors. The aggregator, in order to predict the power consumption and the flexibility of its customer portfolio must classify into categories the consumers. This means knowing how to divide into groups based on consumption behaviours and flexibility in which they operate.

In this way the aggregator is able to obtain clusters of customers.

The process described is in most parts automated and the core is the Energy Box, a smart domestic device that receives various inputs. The latter are weather forecasts, consumption data, interface counters and signals of the aggregator.

The Energy Box needs to know to take into account the preferences of the consumer. This means that the consumer can decide to exclude it on days when he does not want to be disturbed, because they should be free to define their own preferences based on habits.

The counter data are essential to know the profile of the customer.

The aggregator needs to know the total consumption of the consumer, the habits and the load profile. Once the consumer becomes a part of the aggregator's portfolio, the aggregator must make an assessment about its potential flexibility in shifting and modifying the demand.

The birth of so-called buying groups also in this field could change the perspective again 'bottom up'.

Infact, in this view of guaranteeing consumers protection, the aggregator should be an actor that coordinates the offers of all consumers in their exclusive interest, proposing an aggregate load profile and attaining such as it happens in other goods purchasing through FTO, without intermediation through third parties, driving down the final price of the product.

The electrical energy market may occur in a different way:
- bilateral contracts where the distributor enters into a bilateral contract with the aggregator;
- business to business relationships;
- contract platform with different aggregators;
- market that is open to different buyers, which can be both regulated and unregulated use aggregators to improve the profit in the purchase of energy in the energy market.

Another imperative of the intelligent city is the integration that must take place for all urban functions, in a transversal way, from instrumentations to applications, from knowledge to skills, that today more than ever are required to be integrated.

Especially for the most vulnerable contexts it still is an uphill climb, but most cities have for years taken this virtuous path, where skills, systems, and instrumentation contexts, are able to interact continuously to experience innovation in cities.

The integration of mobility systems, on one hand, and integration of systems for the production of energy, on the other hand, are the pillars of the city of the future, focused on a low emissions model and on widespread use of renewable energy. It does not matter whether this is called smart city. The projects[7] undertaken in the Italian cities thanks to specific loans are the evidence of this, as it will be shown in this paragraph.

The ICT technology allows to step up to make sure that certain processes are more immediate.

All the examples of eco-cities, green models, sustainable cities, rely on integrated mobility systems: here the technology plays a key role. The public transport system should be favoured compared to the private sector, because only proposing

---

[7] Some experimental research projects undertaken in Italy since 2012 were selected under the call Miur Smart Cities and Communities, dedicated to the Italian regions suffering from a delayed development and where the Gross Regional Product is below 0.75 times the European average.

to the city several and integrated public transportation systems, a virtuous cycle can be drawn, able to reverse the trend and affect the level of emissions, Freiburg, Stockholm are an example for all cities. From slow to fast mobility, based on non-polluting public transport or transport systems (tram and metro) by rail and not by road. Small and big cities, but with highly integrated systems of transportation.

In Stockholm, such as in Freiburg the slow mobility, the bike is one of the most used transportation mode. There are, in both cities, kilometres and kilometres of bike paths running through the city. It is so that—in the streets of the city centre there is not much traffic and cars use is very limited. The description of the municipal car park of the city of Stockholm allows making an initial evaluation: 1,600 Buses, for 40 %power is drawn from biofuel (ethanol and biogas).

While 15 % of private transport use biofuel, 50 % employ ethanol and diesel. And it is quite advanced a project carried out by the City of Stockholm to the increase the use of electric vehicles (Environmentally Friendly Vehicles).

Even for the energy sector, projects, experiences, and more recent studies show that there is a need to work on different systems of energy production to satisfy the demand in a clean way.

One example is a simulation carried out for the island of Pantelleria in the Mediterranean Sea in the context of a master thesis developed at the University of Palermo (Riva Sanseverino et al. 2014). The preliminary study of the territory of Pantelleria showed a rather marked growth in consumption, in a context of energy supply based almost entirely on fossil fuels. The local energy production from renewable sources is still too limited, though the territory presents interesting potential for exploitation that should be properly analysed and developed. The island is rich of renewable energy resources (solar, wind and geothermal), which led to design three different integrated scenarios (solar, wind and geothermal) that can bring the island to be almost independent from the energy point of view.

The study carries out a preliminary analysis of the island from the point of view it is of environmental constraints exist, the characteristics of settlement, statistics and cultural landscapes, in order to mitigate the possible impact of new energy systems in the area.

In the energy field the recent Project Res Novae[8] (Project Res Novae 2013) proposes the design of a platform for the optimized management of the energy flows at urban level, based on the use and on the integration of innovative enabling technologies for the development of smart grids. Recently (September 2013) in Bari, the mayor and a representative of ENEL have recently presented the digital platform Res Novae (acronym of Reti, Edifici, Strade, Nuovi Obiettivi Virtuosi per l'Ambiente e l'Energia).

The project led by ENEL is also carried out by General Electrics, IBM, some small medium local enterprises, the Italian national council of research CNR, ENEA (another national research centre), the Polytechnic university of Bari and University of Calabria. The project which involves two densely populated urban

---

[8] Among these, the project Res Novae, which offers a platform for the integration of innovative technologies for smart grids.

areas of the southern Italy (Apulia and Calabria), as the cities of Bari and Cosenza, will be developed over a period of about two years, and tries to set one of the most important parts of the puzzle that aims to achieve a sustainable urban environment. The City of Bari is indeed working since 2011 to be a "Smart City", competing with other European cities; the administration indeed signed the "Covenant of Mayors", active on issues of environmental sustainability. Bari after having opened last July, the first network infrastructure of the South dedicated to electric vehicles recharging, is now betting on Res Novae (Project Res Novae 2012), with an investment of around € 24 million, co-financed by the Ministry of Education, University and Research.

The research will lead to the implementation in Bari of a "Urban Control Centre", which will provide information about the urban environment in the government, citizens and other stakeholders. A sort of control room of the city, where the Municipal government will have the necessary information about energy to dynamically support the actions for strategic regional planning.

References

AGCOM (2012) Osservatorio trimestrale sulle Comunicazioni

Caragliu A, Del Bo C, Nijkamp P (2009) Smart Cities in Europe. Paper presented at CERS'2009-3rd Central European Conference on Regional Science, Košice,SlovakRepulic,Oct. 2009, http://www.cers.tuke.sk/cers2009/PDF/01_03_Nijkamp.pdf

Cassa depositi e prestiti (2013) Smart City. Progetti di sviluppo e strumenti di finanziamento

Cittalia—fondazione ANCI ricerche (2012). SmartCities nel mondo

Clancy H (2013) Smart city spending to reach $20 billion by 2020. http://www.greenbiz.com/news/2013/03/06/growth-smart-cities?page=0%2C0

Correia G, Viegas JM (2009) A conceptual model for carpooling systems simulation. J Simul 3:61–68

European Commission (2011) Roadmap to a single European transport area—towards a competitive and resource efficient transport system, white paper

European Commission (2012) Smart grid projects in Europe: lessons learned and current developments

Evans D (2011) The internet of things—how the next evolution of the internet is changing everything, white paper, Cisco

Expert Working Group on Smart Cities Applications and Requirements (2011) Net!Works European Technology Platform, White Paper 20 May 2011, http://www.networks-etp.eu/fileadmin/user_upload/Publications/Position_White_Papers/White_Paper_Smart_Cities_Applications.pdf

Giffinger R, Fertner C, Kramar H et al (2007) Smart Cities—Ranking of European medium-sized cties, research report, University of Technology, Vienna, http://www.smart-cities.eu/download/smart_cities_final_report.pdf

Green J (2011) Rethinking urban renewal. http://dirt.asla.org/2011/11/23/rethinking-urban-renewal/

Lamborghini B, Donadei S (2006) Innovazione e creatività nell'era digitale. Le nuove opportunità della Digital Sharing Economy, Milano Franco Angeli

Mattei U (2011) Beni Comuni—Un manifesto, Laterza, Roma-Bari

Montagnini E, Reggiani T (2010) Nuove Forme di Consumo e Socializzazione: I Gruppi di Acquisto Solidale (GAS), in Consumatori, Diritti e Mercatlo, 1/2010, pp. 91–101 (on-line)

Orlando S (2012) Le informazioni, Cedam, Padova

Petroni P (2011) Il Progetto ADDRESS—TdE 13 ottobre 2011—Sintesi degli interventi, Milano Politecnico

Project Res Novae (2013) Il futuro dell'energia parte da Bari, pubbl. il 24 Sept 2013 in www.enel. it/il-IT/even/ti_news/

Riva Sanseverino E, Riva Sanseverino R , Favuzza S, Vaccaro V (2014) Near zero energy island in the mediterranean. Energy Policy 66

Rossi A, Brunori G (2011), Le pratiche di consumo alimentare come fattori di cambiamento. Il caso dei Gruppi di Acquisto Solidale., Agriregionieuropa, 27/2011 (on-line)

Saroldi A (2008) Gas (Gruppi di acquisto solidale), in Aggiornamenti Sociali, 1/2008, pp. 65–68 (on-line)

Saroldi A (2001) Gruppi di acquisto solidali. EMI, Bologna

Time Plan and Schedules Project (2005) City of Bolzano, http://www.comune.bolzano.it/index_ it.html

Valera L (2005) GAS Gruppi di Acquisto Solidale. Edizioni Terre di Mezzo, Milano

WiSafeCar Project (2013). http://www.wisafecar.com

Urban Smartness: Tools and Experiences

Domenico Costantino

Abstract

The necessary steps to build a different city that combines both sustainable development and urban quality include understanding of the events that emerge in different territories, identifying the appropriate actions, policies and finding innovative tools and procedures.

4.1 The unsustainable settlement model

The growth of urban territories and the concentration of the population in those urban areas characterize the current development of the settlements.

Beginning with the second half of the twentieth century, the number of inhabitants populating urban areas has grown exponentially: in fact, in the second half of '900 only 30 % of the world population lived in cities, while, nowadays, about 50 % of the population, accounting for 3.3 billion, are city residents. The United Nations in "The State of World Population 2013" report that this phenomenon will continue to increase in the next decades, and it has been calculated that, by 2050, 70 % of the world population will live in cities (UN-Habitat 2013).

Analysing satellite information gathered for twenty years (1988–2008), researchers in the American universities have published an article in PLoS ONE (Seto et al. 2011). The data they have analysed in this paper show that

D. Costantino (✉)
Department of Architecture, University of Palermo,
Viale delle Scienze, 90128 Palermo, Italy
e-mail: domenico.costantino@unipa.it

E. Riva Sanseverino et al. (eds.), *Smart Rules for Smart Cities*, Sxi 12,
DOI: 10.1007/978-3-319-06422-2_4,
© Springer International Publishing Switzerland 2014

urbanized territories from year 1970 to 2000 amount to about 58,000 km^2. India, China, and Africa have experienced the highest rates of urban land expansion, while the largest change in total urban space has occurred in North America. It is expected that, by the year 2030, the urban expansion will account for about 1.5 million km^2, an area as large as Mongolia.[1] The results of the study reveal that in several Countries the variations of the urban expansion rate correlate to the degree of household income, to the socio-economic situations in the nations and regions in addition to the political situation.

This rapid and intensive urbanization is the most impressive and irreversible landscape reshaping made by man, which undoubtedly will cause environmental changes and global warming. Not to mention the destruction of the ecosystems and of coastal territories, rivers, agricultural lands, forests, savannas and natural habitat areas that are extremely vulnerable and very valuable.

Throughout the world, like in Europe[2], both large megalopolis and small towns are the places where the majority of the population prefer living in because cities are thought to be special and exclusive places, the heart of national and global socio-economic systems. The most competitive cities are characterized by population growth, availability of employment positions, but also by the ability to attract financial, political and cultural activities as well as traffic congestion.

Nevertheless, cities are high energy-consuming places, in fact, they use about 75 % of the world energy supply and contribute significantly to the greenhouse gas emissions (80 % of the gas emissions of CO_2), entailing high economic costs like consumption of soil, water and energy, pollution, inefficiency of the productive structures and services, shortage of housing, urban decay). Moreover, cities are characterized by serious social disadvantages like high unemployment rates, job insecurities and social marginalization, spatial segregation, poverty, crime, migration.

Cities, in the other parts of the world, are built following the models, architecture and shape of western cities that, at international level, are considered models of modern culture, but plagued by an unsustainable growth rate and the need of considerable amount of resources.

Cities have a very important role in climate change because negative effects are considerably greater, and global problems are amplified when combined with local phenomena (microclimate, pollution, etc.). This combination creates events like the "heat island".

The layout of the city greatly influences the urban microclimate because of extensive overbuilding, decreasing in agricultural areas, soil sealing, expansion of paved areas compared to green spaces. Moreover, physical structure, urban areas

---

[1] Analysis has been developed by K. C. Seto, M. Fragkias, B. Güneralp, M. K. Reilly. The study represents the first estimate of how urban areas are growing globally and how they can grow in the future. Results show considerable variation in the rates of urban expansion over the study period, with the highest rates in China followed closely by Southwest Asia. Average rates of urban expansion are lowest for Europe, North America, and Oceania.

[2] The 73 % of Europe population lives in urban areas, produces around 80 % of GDP and consumes up to 70 % of the energy.

and architectural volumes are made of materials (asphalt, concrete, bricks, cement, etc.), which keep heat in and don't allow evaporation. In addition, the geometrical shapes, that characterize the urban plans, create urban canyons that don't allow heat dispersion and shape the wind circulation patterns.

Factors like water wasting and transformation of the hydro-geological system (artificial diversion of water courses, colonization of watery areas) need to be taken into consideration. The community processes that alter the ecosystem and limit the natural course of resource renewal also contribute to alter the environmental conditions. Extreme weather events expose the inhabitants to serious risks or very dangerous situations in specific seasons of the year. Cities, which are located along coastal areas, are more exposed to the unpredictable effects of the sea level changes

The crisis of the settlement model, though, is not limited to extraordinary natural disasters or to health and humanitarian calamities, includes ordinary living habits, production, mobility, free time all considered as factors of risk and decay.

In today's modern society the environmental, socio-economic and urban crises overlap and inter-mixforcing the society into a profound and radical re-thinking of settlement structures and shape in order to increase decarbonisation of urban economy. The replacement of high carbon with low carbon containing fuels will contribute to reduce resource consumptions (soil, water, energy, etc.), emissions of dangerous gases, production of liquid and solid waste and urban sprawl.

Nowadays, we wonder which road cities should take to tackle worldwide problems created by climate changes, financial crises, global economic market, increase of urban and metropolitan competitiveness. Pollution and waste, at local level, are the main causes of soil consumption and destruction of the environmental and landscape resources.

The solutions taken into consideration lead us to critically reconsider the well-established standards of urban planning and territory management, to reassess the deep-rooted beliefs and interpretative patterns, not to mention the methods and the strategies of well-tested interventions that now appear insufficient and inadequate to face the problems of the contemporary metropolis (Gabellini 2013).

In city planning, it is particularly important to re-consider and to find new tools and ways of drafting out the new project in terms of both content and procedures.

It clearly appears that climate change and sustainability of the urban systems have as central issue the integration of climate policies with those of territory management.

Cities do not just represent places where people live with problems but they also offer opportunities and changes. Cities are places where the inhabitants can find extraordinary chances to innovate technology and society. Cities can be considered labs where people experiment new development models (knowledge and culture, research and diffusion of new technologies, green economy, etc.). Innovation, connected to the knowledge and culture activities, depends mostly on urban external economies[3]. Innovation is more and more a social structure that occurs in an urban context plenty of opportunities, interactions, qualified cooperation and information exchange. Cities, as multicultural centres of global communication, have the power to change the world and to write the agenda for a sustainable future.

4.2   Sustainability Proposals

Under the Kyoto Protocol, the EU has set strategies and has enacted directives to ease the transition towards a low carbon economy and improve the cities' innovation potentials. Reducing the greenhouse gas emissions of CO_2, preserving the entropic stability of the global ecosystems are the main intervention policies to be implemented. It was also proposed to modify the production techniques and processes and, at last, to change individual and collective behaviour. The EU has set its own targets by 2020: CO_2 emissions must be cut down by 20 %, the energy efficiency and the energy production from renewable sources should be increased by 20 %, and that of the biofuel for vehicles by 10 % (EU Commission 2010).

Europe 2020 is the ten-year growth strategy of the European Union. It is more than just overcoming the crisis, which unfortunately still afflicts most of our economies. It is about tackling the deficiencies of our growth models and creating conditions for a different type of growth that is smarter, more sustainable and more inclusive. On December 2011, the European Commission adopted the Communication "Energy Roadmap 2050" (EU Commission 2011). The EU is committed to reducing greenhouse gas emissions 80–95 % below 1990 levels by 2050. In the "Energy Roadmap 2050" the Commission explores the challenges posed by the decarbonisation objective, while at the same time ensuring energy supply and competitiveness (ECF 2010; EEA 2012).

These are the main goals of the Climate and Energy Packages written in the EU White Paper "Adapting to climate change: Towards a European framework for action" (EU Commission 2009) which sets out the basis for an European adaptation strategy to reduce vulnerability and correctly tackle the consequences of climate change and strengthen European resilience.

The goal 20-20-20 is an integrated approach to climate and energy policy aiming at fighting climatic changes, increasing EU energy security and at strengthening our own competitiveness.

In April 2013, the European Commission adopted a climate change adaptation strategy. The main document "An EU Strategy on adaptation to climate change" defines the targets and actions to be taken, in order to make Europe more resilient EU Commission 2013).

---

[3] The main activities of the knowledge economy are: high technology, professional services to businesses and individuals, the cultural industry (media, television, cinema, music, cultural tourism), but also productions neo-craft (fashion, design, etc.), services able to answer a question very specialized.

The strategy meets the EU general political and economic targets indicated in the transition towards a sustainable economy, an efficient use of the resources, the respect of ecology and the implementation of low carbon emissions.

The main objective of the EU Strategy on adaptation is to make Europe more climate-resilient. This requires a better preparation and a greater ability to react to present and future impacts caused by climate changes. All levels of government (local, regional, national and of the EU) should be involved in order to achieve a coherent and coordinated approach.

The strategy focuses on three main objectives:

- Promoting action by Member States. The Commission recommends that all Member States adopt national adaptation strategies. The EU will provide financial support for adaptation through the proposed LIFE instrument which includes a climate action sub-programme. The Commission will support the exchange of good practice among Member States, regions, cities and other stakeholders. Building upon the success of its pilot project "Adaptation strategies for European cities", the Commission will continue to promote urban adaptation strategies. Adaptation action by cities will, in particular, be developed in coordination with other EU policies following the model of the Covenant of Mayors.
- Better informed decision-making. The Commission will fill the gaps in the knowledge of Climate Change adaptation. The future EU programme for research and innovation, Horizon 2020, will address climate change adaptation through its "societal challenges" priority investing in excellent science and promoting innovation. It will be given greater impetus to the European Platform on Climate Change "Climate-ADAPT" with better access to information and greater interaction with other platforms.
- Climate-proofing the Commission will promote adaptation in vulnerable key sectors. One priority and responsibility for the Commission is to mainstream adaptation through legislations in such sectors as marine waters, forestry, and transport and in important policy instruments such as inland water, biodiversity, migration and mobility. The Commission will continue its work to facilitate the climate proofing of the Common Agricultural Policy (CAP), the Cohesion Policy and the Common Fisheries Policy (CFP). The Commission will ensure that Europe can rely on more resilient infrastructures through a review of energy, transport and construction standards. In addition, it will promote insurance and other financial products for resilient investments, business decisions to reduce risks in the European market and measures into EU policies and programmes.

The National Adaptation Strategies (NAS) are at different stages of planning and implementation. Sixteen Member States have formally adopted their own strategy of adaptation at this time, while twelve more Member States are already in an advanced phase of adaptation. The Netherlands, for example, have already implemented an important climate change adaptation Plan, combining different expertise and involving competence on water integrated management with networks and localization of new settlements[4].

Germany and the United Kingdom have already successfully started assessment and monitoring of the adaptation plan using environmental indicators.

Urban and Regional Adaptation Policy Measures are still to be defined (like in several British counties: Staffordshire, South East).

Several cities are making specific plans to solve problems caused by climate change (Birmingham, Madrid, Stuttgart, Sfantu Gheorghe, San Sebastian) while Rotterdam, Copenhagen and Aalborg are examples of smart adaptation strategy (EU, DG Clima 2013).

Even Italy is creating a new adaptation policy, since it doesn't have appropriate existing measures on the subject (environmental preservation, prevention of natural disasters, sustainable management of natural resources and health preservation).

The Italian Ministry of Environment, awaiting for the National adaptation strategy, in 2013 published the "Public consultation document" containing the guidelines to collect the suggestions of the stakeholders and to promote cooperation agreements (MATTM 2013).

The main objective of the national adaptation strategy is to develop a national perspective on how to face future impacts on climatic change, to identify actions and methods to tackle climate variability and extreme weather conditions. By implementing actions it will be possible to reduce to the minimum the risks arising from climate changes.

Other objectives are the protection of health, welfare and goods of the population, the preservation of the natural asset, the maintenance and improvement of the adaptation capacity of the natural and socio-economic systems. The implementation of the adaptation plans will potentially offer advantages and opportunities. The document was written taking into account the general and already well-established experiences in other European countries like Belgium, Denmark, Finland, France, Germany, the United Kingdom, Spain and Switzerland. Furthermore, the reports and technical articles of the European Environment Agency (EEA) were taken into consideration.

As far as city policies are concerned, three relevant actions were enforced in the last few years: the creation of the Inter-Ministerial Committee for Urban Policy (CIPU), "Plan Cities" for the requalification of degraded urban areas, predisposed by the Ministry of Infrastructure and Transport and the project of national interest "Smart Cities" funded by the Ministry of Education and Research in collaboration with university research groups.

---

[4] Since 2005, sixteen Member States have formally adopted strategy of adaptation: Finland (2005), Spain (2006), France (2007), Hungary (2008), Denmark (2008), The Netherlands (2008), UK (2008), Germany (2008), Sweden (2009), Belgium (2010), Portugal (2010), Switzerland (2012), Malta (2012), Ireland (2012), Austria (2012), Lithuania (2012).Twelve other states are at an advanced stage adoption of a strategy: Bulgaria, Cyprus, Czech Republic, Estonia, Greece, Italy, Latvia, Norway, Poland, Romania, Slovakia, Slovenia.

The legislator, appointing the Inter-Ministerial Committee for Urban Policy (art. 12Bis law 134/2012), wanted to overcome a disjointed and limited management of urban areas problems[5]. The CIPU is an important institutional innovation since it recognizes that cities are central to economic development, social inclusion and territorial cohesion.

It tries to fill the gap between the importance of cities and the lack of appropriate city measures identified by the national government.

The CIPU has two essential objectives:

- to have a role in coordinating important national intervention actions and be a reference point for local government whose representatives are also part of the Committee;
- to set the Urban Agenda, in accordance with the one suggested by the European Commission concerning the cohesion policy 2014/2020. Annual programming and budget reports will periodically update this agenda.

The Urban Agenda is a policy document that the majority of the European countries have already established, with the indication of the priorities and coordinated actions at different levels of the Government. Moreover, the agenda fosters sustainable and inclusive investments to strengthen the role of the cities. Metropolitan cities will have a leading role in the new planning of structural funds 2014/2020. A new national operational program specific for Metropolitan cities will be proposed because within the European Regional Development Fund (ERDF) is estimated that at least 5 % of allocated national resources should be assigned to Integrated Actions for a sustainable urban development.

CIPU took office on January 2013. The Ministry for territorial cohesion, during the second meeting presented the document "Priority methods and contents concerning Urban Agenda" in which the bases, the methodology, the contents, the priorities to be followed for the Urban Agenda were indicated. The text was written taking into account the different urban dimensions (metropolitan areas, large and medium size cities and systems of municipalities) and specific issues present in Southern Italy and in the "In-land Areas". The text underlines the strength, the weakness and the operational priorities of urban systems.

The Urban Agenda contains suggestions to integrate different levels of management, sectorial policies and financial resources coming from the ordinary budget or the European Community funds (CIPU 2013). The Urban Agenda should anticipate more appropriate actions, means and adequate places to solve disparities and problems, to find possible solutions, to overcome and to contrast ordinary and extraordinary policies that, up to now, have replaced the lack of ordinary policies on the cities.

---

[5] In 1987 was established the Ministry for Urban Areas, which operated until 1993, and the Department for Urban Areas at the Presidency of the Council (experience then died out). Subsequently, some ministries have addressed the problems of urban areas: the Ministry of Infrastructure to Infrastructure, the Ministry of Interior to Security, the Department for Development and Cohesion policies to management of European regional policies (in particular, the program Urban).

The "Plan Cities" (art. 12 law 134/2012) expects that Municipalities send to the Ministry of Infrastructure and Transport proposals of "Contracts of urban enhancement" in which they ask for coordinated interventions on degraded urban areas.

At the expiring date (12 October 2012) 457 projects were presented. The Committee "Cabina di Regia", appointed by Ministerial act in August 2012, has evaluated and selected the projects according to the following criteria:

- the operational status of intervention;
- the ability and modality of stakeholders to partecipate and acquire public and private equity funds;
- the reduction of the marginalization and social decay;
- the improvement of the infrastructures;
- the improvement of urban, social and environmental quality.

In January 2013 the list of the 28 selected projects was published[6], allowing a state funding of 318 million Euros, with this financial resources it will possible to realize projects and works worth 4.4 billion Euros coming from public and private investments (MIT 2013). At the same time 24 more projects were evaluated qualified for funding. So far only three cities (Potenza, Matera and Venezia) have signed the "Contracts of urban enhancement" with the "Cabina di Regia" that has the task for monitoring the progress of the project.

The results of the "Plan Cities" will be compared to the previous experiences of urban planning as "Recover Program", "Urban re-qualification Programs" and "District contracts".

4.3   Urban Regeneration

The attention to the problem of resources, compared to the past years, requires a different approach towards urban transformation and requalification proposals.

A greater sense of responsibility, a different moderation and new austerity steps are needed in order to find solutions that are readily available (Oliva 2011).

Regeneration offers an opportunity for environmental, architectural and social requalification of cities.

This is a different approach compared to the past, since it is an alternative to urban expansion, to building new shapeless urban structures, and to underuse the existing building heritage. This approach isn't restricted to adaptation of cities to climate change, but it focuses, not only on the quality of urban living and citizens participation, but also on the strategies adopted in building urban spaces that will

---

[6] The selected cities are: Ancona, Bari, Bologna, Catania, Cagliari, Eboli, Erice, Firenze, Foligno, Genova, L'Aquila, Lamezia Terme, Lecce, Matera, Milano, Napoli, Pavia, Pieve Emanuele, Potenza, Reggio Emilia, Rimini, Roma, Settimo Torinese, Taranto, Torino, Trieste, Venezia, Verona.

result in reduction of vulnerability and increase of resilience[7]. These perspectives offer alternative patterns of interventions with a potentially revolutionary impact on cities, on government structures, on professional organizations, on formative processes and on research.

Urban regeneration should be interpreted as inter-sectorial and integrated policies for sustainable development, and it should be considered strategic for local development, aiming at promoting and enhancing the urban systems with the support of the State in partnership with private investors.

Urban quality, revisited according to the concepts of regeneration and resilience, needs a sustained process of preservation and creative interventions to overcome stress factors and environmental pressure (endogenous/external) generated by ordinary events and by geological, weather and socio-ecological emergencies.

Several countries are already planning and implementing adaptation strategies for building structures (eco-buildings), networks (sustainable mobility, smart grid), settlements (eco-districts, smart cities), and experimenting new ideas of cities more sustainable and pleasant to live in (Costantino 2012). It is now clear, how energy saving and use of renewable sources are a great opportunity to promote economic development and independence from conventional energy sources.

Local authorities understand the strategic importance of these actions and now they use their managerial expertise in planning development in this direction. Local governments play a leading role in meeting the climate and energy targets set by the EU.

The "Covenant of Mayors" is an agreement in which villages, cities and regions voluntarily commit themselves to endorse the European Union's target of 20 % CO_2 reduction (Covenant of Mayors 2014). This commitment must be pursued by implementing the Sustainable Energy Action Plan (SEAP). The Covenant of Mayors is a bottom-up support, being the only movement with the capacity to mobilize local and regional authorities to satisfy the European objectives. Its role is also of support for efforts made by local governments to implement energy saving policies.

Even the global network of large metropolis "Cities Climate Leadership Group" (C40 2014) has the objective to promote development and to implement policies and programs for the reduction of greenhouse gas emissions, climate risks and for the increase of energy efficiency in large cities around the world[8].

The first edition of the "City Climate Leadership Awards" sponsored by Siemens and C40 (September 2013) prized the first ten cities, recognizing them as leaders in supporting urban sustainability and in fighting climate change. The prize

---

[7] Resilience is understood not only as ability to withstand an external stress, but also to be able to creating new and better conditions because "Resilience is the ability to continue to exist, incorporating the change" (Berkes et al. 2008).

[8] The Cities Climate Leadership Group C40 is a network of the world's megacities taking action to reduce greenhouse gas emissions. It was created in 2005 by the Mayor of London Ken Livingstone. C40 harnesses the assets of member cities to address climate risks and impacts locally and globally. C40 now has grown to 63 members.

is a worldwide recognition given to those cities whose performances and competences, in the protection of climate change, are considered as referral points for all the other cities. The winning cities are: Bogotà, for "Urban Transportation" with its "TransMilenio" project on electric and hybrid buses and taxis; Copenhagen for "Carbon Measurement and Planning" for its "2025 Climate Plan" which outlines the guidelines that the city intends to follow to become the first carbon neutral capital city by 2025; Melbourne for the "Energy Efficient Built Environment" for its "Sustainable Buildings Program" improving the efficiency of Melbourne's commercial buildings; Mexico City for the "Air Quality" with its "ProAire" program whose implementation has made possible to accomplish, over the last two decades, an impressive reduction on air pollution as well as in CO_2 emissions was obtained; Munich for "Green Energy" with its "100 % Green Power" plan whose standards and requirements needs to be met by 2025; New York City for "Adaptation and Resilience" with its "A Stronger and More Resilient New York" which will concentrate on the reconstruction of the communities hit by hurricane Sandy and on more resilient structures and buildings of the city; Rio de Janeiro for "Sustainable Communities" with its "Morar Carioca" project, a comprehensive urban revitalization strategy which will invest in revitalization projects throughout the city with the aim at formalizing all the city's favelas by 2020; San Francisco for "Waste Management" with its "Zero Waste Program" which started in 2002; Singapore for "Intelligent City Infrastructure" with its "Intelligent Transport System" plan and Tokyo for "Finance and Economic Development" with its "City Cap-and-Trade Program", launched in 2010, which intends to considerably reduce CO_2 emissions by large commercial, industrial and government buildings.

Knowing other cities' experiences is an opportunity to analyse public policies and to change already built cities, it is also a good occasion to compare the application of different cultural approaches and different local policies during urban regeneration.

There are many examples of good practices in urban regeneration that we could describe, but we will list just few of them: Hammerby district, in the outskirts of Stockholm, Vauban in Freiburg, Augustenborg in Malmo, the district of Confluence in Lyon, the vast re-planning program of degraded industrial areas in Ile de Nantes and High Line in New York City.

In Italy, a proposal to start regeneration programs on existing housing stock is the "National Plan for Sustainable Urban Regeneration" (Ri.U.So.) promoted by the National Council of Architects, Planners, Landscapers, Conservationists (CNAPPC 2012), by the National Association of Building Contractors (ANCE) and by Lega Ambiente (Italian Association for the protection of the Environment).

With Ri.U.So. the construction professionals suggest ideas concerning cultural, economic and social transformation and valorisation of the cities. The National Plan for Sustainable Urban Regeneration aims at offering efficiency, safety and quality to 100 Italian cities that host 67 % of the Italian population and are considered as the cultural and productive assets of the country accounting for 80 % of the nation's Gross National Product (GNP).

The proposal is based on two simple but important observations: the depletion of energy resources and the bad conditions of the housing built[9] in Italy after the World War 2.

The objectives are:

- maintenance, regeneration and safety of public and private housing taking into account that more than 24 million people live in seismic hazard areas, while another 6 million people are living in areas with high hydrogeological risk;
- drastic decrease in the consumption of land and reduction of energy wastage and water consumption in housing by promoting the realization of energy and ecological districts;
- revaluation of the public spaces, urban green areas and district services;
- rationalization of urban mobility and waste cycle;
- implementation of innovative digital infrastructures connecting Italian cities, promoting home working, reducing commuting, transferring and wastage;
- preservation and revitalization of historic centres.

To implement the National Plan for Sustainable Urban Regeneration (Ri.U.So) a national policy on the regeneration of cities is needed, giving directions that will ensure quality standards, low costs, minimal environmental impact and energy saving. The Plan also includes a cultural improvement involving all stakeholders, from designers to companies, involved in the building industry.

To achieve these objectives, it is not sufficient the synergy among politics, experts, companies and finance, but the citizens cooperation is also necessary, not only because the cities' housing stock is mainly owned by private subjects but because it is also important that citizens become aware of the quality of the environment they live in.

Italy needs to set politics and tools for urban regeneration, for bridging the gap between what is done in building for sustainability and what can be done for ruling a smart planning of the cities.

In Italy, the city planning has proved to be, in general, incapable to predict and implement adaptation and mitigation actions, to face natural emergencies and to solve the social and economic issues of the city.

Although the "strategic" model was introduced in the administrative and professional practice, the plan has not been able to give concrete answers to the questions emerging from society.

To achieve a new urban quality for the Italian cities it is necessary to integrate the planning with the designing. It is necessary to bear in mind the Italian architectural culture and how this culture grew and developed.

Urban Building Regulations could be the potential connection tool to solve the problems among sustainable building and designing of urban open space including the issues on single buildings and public space.

---

[9] In Italy 70 % of the buildings are over 40 years old and was built with materials and techniques that can be considered obsolete.

The urban Planning of open spaces according to the sustainability criteria allows improvements on climate and on quality of urban life.

In France, for example, "guides" and "fiches" (technical manuals) are widespread guidelines indicating the best practices for building and designing public or private open spaces as well as already constructed areas, like the "Guide de qualité urbaine et d'aménagement durable de la communauté urbaine de Bordeaux" (CUB 2008).

In some regional laws (i.e. Emilia Romagna, Tuscany), the Urban Building Regulations rule on all possible interventions on already constructed areas and on the organization and utilization of open spaces. It contains also the criteria for planning new open spaces and building areas (BR Bologna 2009; Trento 2013). In other regions, the Building Regulations regard just binding rules and technical modality for construction.

Nevertheless, the several European Directives, issued in 2002, concerning energy saving plans, have indirectly modified building regulations that now have a new "Energy and Environmental Annex".

The Annex is, in fact, the main tool used by the local administrations to regulate building transformations according to criteria of environmental sustainability and energy efficiency, to save and rationalize the use of resources, to increase coherence among the interventions, the territory and the Sustainable Energy Action Plan (SEAP).

The Building Regulations, therefore, can be an essential tool to modernize the building sector, promoting people health, protecting environment and saving resources. In the Regulations, technical and procedural aspects and economic interests merge together, while different administrative competences (planning, building and energy) at different levels of government (State, county, municipality) come across.

In 2012, according to the ONRE report 2013, approximately 1003 (12.4 % of total) are the Italian municipalities that have amended and reformulated the building regulations according to criteria of environmental sustainability and energy efficiency. They are big cities and small towns where a total of more than 21 million people they live in (ONRE 2013).

4.4 Conclusions

The contemporary city is a strategic resource for sustainable development; it is also a place with a high concentration of serious problems, and events that contribute to environmental degradation and global warming. The future must start from cities.

Many experts foresee that it is possible to carry out a reversal of trend by relying on technology, revising society organizations and lifestyles.

The concepts of adaptation, mitigation, resilience, urban regeneration, decrease and "shrinking" are largely recurring in the present debate but they are to be experimented and to be verified in the different geographical, political, socio-economic, environmental and cultural contexts.

Nevertheless, it is now clear that sustainable development goes through the cities and that the future is linked to the city's ability to refurbish itself from the inside by adapting to the lack of resources, activating resilience processes, promoting requalification, renovation or radical substitution of existing areas, ensuring quality standards, low costs, minimal environmental impact and resource saving.

It is apparent that, although cities have acquired in-depth knowledge and many experiences have been carried out, the lack of political decisions and the lack of a widespread culture of the urban and environmental crises create in important problems in the realization of a sustainable city.

One of the main causes of the urban crisis is the lack of congruence between the real size of the city and the decision-making model used to manage it. The real problems of cities and territories are not appropriately tackled, but most of the resources available are wasted.

It seems clear that the relationship between society and environment will be more and more strategic for the future of cities and that smart growth will be increasingly related to available resources, knowledge, cultural heritage, social contexts and green economy.

References

Berkes F, Colding J, Folke C (2008) Navigating social-ecological systems: building resilience for complexity and change. Cambridge University Press, Cambridge

Cities Climate Leadership Group, C40 (2014), www.c40.org/

Comune di Bologna (2009) Regolamento urbanistico edilizio. www.urp.comune.bologna.it/PortaleTerritorio/portaleterritorio.nsf/

Comune di Trento (2013) Regolamento edilizio comunale. www.comune.trento.it/Comune/Atti-e-albo-pretorio/Regolamenti/B02-Regolamento-edilizio-comunale

Communauité Urbaine de Bordeaux, La CUB (2008) Guide de qualité urbaine et d'aménagement durable. www.lacub.fr/sites/default/files/PDF/publications/guides/guide_qualite_urbaine.pdf

Consiglio Nazionale Architetti Pianificatori Paesaggisti e Conservatori, CNAPPC (2012) Piano Nazionale per la Rigenerazione Urbana Sostenibile. www.awn.it/AWN/Engine/ RAServeFile.php/f/ Documenti%20CNAPPC/4

Comitato Interministeriale per le Politiche Pubbliche—CIPU (2013) Metodi e contenuti sulle priorità in tema di Agenda Urbana, Allegati 1 e 2, Roma

Costantino D (2012) Verso la smart city, in Riva Sanseverino E, Riva Sanseverino R, Vaccaro V, Atlante delle smart cities: modelli di sviluppo sostenibile per città e territori, Franco Angeli/Urbanistica

Covenant of Mayors (2014), www.covenantofmayors.eu/index_en.html

European Commission (2013) An EU Strategy on adaptation to climate change. www.eur-lex.europa.eu/LexUriServ/LexUriServ.do?uri=COM:2013:0216:FIN:EN:PDF

European Commission (2011) Directorate General for Regional Policy Cities of tomorrow—Challenges, visions, ways forward. www.ec.europa.eu/regional_policy/conferences/ citie-softomorrow/index_en.cfm

European Commission (2010) Europe strategy 2020, a strategy for smart, sustainable and inclusive growth. www.ec.europe.eu/europe2020/index_en.htm

European Commission (2009) White Paper, Adapting to climate change: towards a European framework for action. www.eur-lex.europa.eu/LexUriServ/LexUriServ.do?uri=COM:2009:0147:FIN:en:PDF

European Commission DG CLIMA (2013) Adaptation strategies for European cities. www.eucities-adapt.eu/cms/

European Climate Foundation, ECF (2010) Roadmap 2050: A practical guide to a prosperous, low-carbon Europe. www.roadmap2050.eu/attachments/files/Volume1ExecutiveSummary.pdf

EEA (2012) Urban adaptation to climate change in Europe-Report n.2. www.eea.europa.eu/publications/urban-adaptation-to-climate-change

Gabellini P (2013) La rigenerazione urbana come resilienza, in XXVIII Congresso Nazionale INU, Salerno 24/26 Ottobre 2013, in www.inu.it/congressi-inu

Ministero Ambiente Tutela Territorio Mare, MATTM (2013) Elementi per una Strategia Nazionale di Adattamento ai Cambiamenti Climatici, Documento per la Consultazione Pubblica. www.minambiente.it/sites/default/files/archivio/comunicati/Conferenza_29_10_2013/Elementi per una Strategia Nazionale di Adattamento ai Cambiamenti Climatici.pdf

Ministero delle Infrastrutture e Trasporti, MIT (2013) Piano Città. www.mit.gov.it/mit/site.php?p=cm&o=vd&id=2404

Oliva F (2011) La città oltre la crisi. Risorse, governo, welfare—Relazione introduttiva, XXVII Congresso Nazionale INU, Livorno 7/9 Aprile 2011, www.inu.it/congressi-inu

Osservatorio Nazionale Regolamenti Edilizi, ONRE (2013) L'Innovazione energetica in edilizia, Rapporto ONRE 2013. www.legambiente.it/sites/default/files/docs/sito_onre_2013_min.pdf

Seto KC, Fragkias M, Güneralp B, Reilly MK (2011) A meta-analysis of global urban land expansion. PLoS ONE information. www.plosone.org/article/info%3Adoi%2F10.1371%2Fjournal.pone.0023777

UN-Habitat (2013) State of the world's cities 2012/2013, prosperity of cities, New York. www.unhabitat.org/

The Urban and Environmental Building Code as Implementation Tool

5

Valentina Vaccaro

Abstract

The frame within which the work is placed refers to the actions necessary to achieve the objectives of the coverage of the consumption of energy from renewable sources compared to the gross final consumption, posed to Regions by 2020, and that can be implemented through various actions involving local governments including the revision of the municipal building codes in a sustainable view. These actions are increasingly being recognized as energy planning tools for the territories where administrations have committed to the European project Covenant of Mayors. The discussion shows how the adoption by the Regions of Guidelines for sustainable municipal building codes can be a practical tool for raising the energy performance of buildings and the achievement of common goals of sustainability at regional scale. The work also aims at showing a concrete example of the definition of guidelines for the revision of the municipal Building Regulations for cities within the Sicilian Region.

V. Vaccaro (✉)
DEIM, University of Palermo, Palermo, Italy
e-mail: valentina.vaccaro03@unipa.it

E. Riva Sanseverino et al. (eds.), *Smart Rules for Smart Cities*, Sxi 12,
DOI: 10.1007/978-3-319-06422-2_5,

5.1   The "Building Regulation" as Implementation Tool for Sustainable Restoration of Existing Buildings

Based on current political guidelines of the European Community that push strongly towards the achievement of objectives aimed at the energy sustainability of cities, of buildings and of facilities, the building code seems to be the best operative instrument to ensure a sustainable regulation of the municipal territory, taking into account the local context to which it refers.

In Italy it is well-established the interest in this instrument, and at the level of municipal planning, it is the document more suitable to introduce new energy and environmental criteria and objectives, which are improvements if compared to existing legislation that regulates the field of energy efficiency and use of plants producing energy from renewable sources in the building sector.

It is interesting to note in Italy the growth, year after year, not only of the number of municipalities involved with a revision of its building code in a sustainable way, +42.3 % compared to 2013 to 2010 and even +80 % compared to the 2009 (Cresme Ricerche s.p.a and Legambiente 2013) but also the features accounted for in it (the use of renewable energy production plants; the water saving; the use of trees to improve external microclimatic conditions; such as also features referred to the coating and the conditioning systems). By now the experiences relate to all regions of Italy, with at least one sustainable building code in each Italian region.

This interest is due to the fact that municipal building code is a focal point of the building process, where political, technical and procedural aspects collide. It must account for political and technical stakeholders' needs.

Besides, cross competencies in the fields of urban administration, planning, construction and energy management are needed for its definition. This document indeed is an operational tool strongly connected to the territory and defined through a 'bottom-up' approach, accounting for critical issues at local level. On the other hand, it must be flexible enough to adapt to the race towards 'urban sustainability' that the European Union is running.

The achievement of targets for reducing climate-altering emissions by 2020 and beyond, and the definition of appropriate means to that end, are elements that characterize the energy policy of the Member States who have a natural implementation of regulatory measures at national, regional and also municipal level with appropriate insights to the various spatial scales to which they refer.

Since 2002, with the first Directive on energy efficiency in buildings, the European Union, in fact, decided to initiate a process more and more complex and an in-depth change that today has been updated with the latest Directive 2012/27/UE on efficiency energy.

In this context, for the achievement of national and regional goals of polluting emissions reduction, it is more and more important the role of local authorities.

Especially in the building sector, which is ruled and controlled at Municipal level, putting in place on the territory a 'bottom up' involvement that now seems to be a shared element in all energy policies aiming at a real change.

5.2 The Revision of Municipal Building Codes in the Context of Regional Policies for Energy in Italy

Recently, many Italian regions have issued regional Guidelines for the definition of rules such as urban planning regulations, aiming at energy and environment valorisation (Dall'O and Galante 2009).

The aim of these Guidelines is to define a univocal set of contents for the updating of municipal building codes from a sustainable perspective.

In this way, the Municipal administration could integrate and implement objectives and requirements aiming at environmental sustainability of buildings, in line with the tools of regional energy planning.

The regulatory framework in which is in the definition of a model of building code for municipalities, issued at the regional level, is related to the actions necessary to achieve the goals by 2020 seats to the regions of Italy and agreed at the national and European level.

The commitments at European level, synthetically represented by the 20-20-20 model, and declined at national level for the different regions by the so called "Burden Sharing", Italian Ministerial Decree March 15th (2012), have given each region, coherently with the National Action Plan,[1] intermediate and final objectives. These objectives refer to the coverage of energy consumption from renewable energy sources as compared to the gross final consumption of thermal and electrical energy as well as the energy used for transportations (such percentage for Sicily is for example equal to 15.9 % by 2020).

Even if the reduction of the gross final energy consumption is not a constraining objective for regional administrations, it is clear that with a reduction of the gross final consumptions, the regions can more easily reach the objective. The portion to be covered using renewable energy sources, indeed, is fixed by the Decree as the ratio between the amount of gross final energy consumptions, to be covered using renewable energy sources, and the gross final energy consumption of the region. If the latter is reduced, the target percentage can be reached more easily.

---

[1] The National Action Plan, Directive 2009/28/CE of the European Parliament and of the Council, April 23rd 2009, based on the incentivation of the use of renewable sources for the energy needs, is a programming document. The latter gives detailed indications about the actions to be set out in order to reach by 2020, the constraining objective for Italy to cover the 17 % of the gross national consumptions through renewable energy. The objective can be reached using renewables sources in the following areas: Electricity, Heating–Cooling and transportations (in http://approfondimenti.gse.it).

Considering that in Italy the energy needs for building are one third of the total consumed energy, it is clear that measures such as the revision of the municipal building regulations, supported at regional level, is a valuable tool for supporting the energy policies of the Italian regions by 2020.

The building code, if suitably revised through energy efficiency and sustainability criteria, can support actions on the building and its technical systems, which in turn produce a reduction of gross final consumptions and an increase of renewable energy systems.

The art. 4 of the Burden Sharing Decree states what tools the regions can use to reach the reduction of the energy consumptions.

In general terms, the regions can:

- develop action plans to improve energy efficiency and the implementation of renewable sources at district and territorial levels;
- integrate the programming about renewable sources plants implementation and energy efficiency measures implementation with other sectorial actions.

More specifically, the activities that can favour the attainment of the posed objectives are reported below:

- measures and actions for local public transportations, in buildings and at consumers sites in the regions and at local level;
- measures and actions to reduce urban traffic;
- measures and actions to reduce the electrical energy consumptions for public lighting and in the water sector;
- widespread of financing tools through third parties of the energy services;
- incentive policies for energy efficiency, in agreement with the national laws.

As specified in art. 3 of the Burden Sharing Decree, the regions, in order to ensure the achievement of these objectives are strongly invited to integrate their own tools for the government of the territory and to support innovation in the productive sectors, with specific new provisions for energy efficiency and use of renewable energy sources.

There are many funding programs and projects that the European Union provides both to public and private bodies, which want to implement projects aiming at environmental protection in various sectors. From territorial governance, in reference to strategic and innovative approaches pointing at environmental sustainability, to communication and information projects for raising public awareness in relation to the problem of climate change and resources use, as well as projects closely related to protection of nature and biodiversity.

In this context, the LIFE+ and the Covenant of Mayors Projects for many regions of Italy, including Sicily, are largely influencing the future regional energy planning strategy towards the achievement of the 2020 targets.

The Covenant of Mayors is a European framework to actively involve cities in the strategy towards energy and environmental sustainability. The initiative, launched in 2008 as part of the second edition of the EU Sustainable Energy Week, led to the adherence of more than 1,600 cities, including 20 European capitals and many towns in non-EU countries, with a mobilization of more than 140 million citizens. The Covenant, at present, is one of the main tools for the development on the national territory of the principles of urban sustainability through an integrated approach to the various sectors (building renovation, sustainable mobility, high-efficiency lighting, citizens awareness raise, etc.), bringing as added value an engine for economic and image recovery of many local communities (Lumicisi 2013).

5.2.1 The Case of the Sicilian Region

The involvement of local political authorities, but also of citizens and business, is of fundamental importance within the energy policies of the Sicilian Region. The Covenant of Mayors is just one of the major tools for a broad involvement of municipalities to reach energy savings targets and to increase renewable energy systems.

Sicily, in fact, is the only region in Italy that is putting in place a funding scheme for municipalities wishing to sign the Covenant of Mayors agreement. The funding is proportional to the number of citizens of the municipality.

The financial support is given for the definition of the Basic Inventory of Emissions (BEI) and the Sustainability Energy Action Plan (SEAP) to reach the reduction of CO_2 emissions of about 20 % as compared to a given reference year. Analysing the SEAP approved by the JRC[2] for Italy, it is evident that more than 50 % of the municipal administrations has included among the actions of their SEAP the modification of the building code, showing that such tool is quite important to improve the energy efficiency of buildings and concretely produce a reduction of CO_2 emissions.

The yearly report about the energy audit of the Sicilian Region (Energy Report 2012, elaborated by the Energy Department of the Sicilian Region) provides evidence about how much the building sector is a central issue in the energy policy of the region.

From the analysis of the gross final energy consumptions in Sicily for years 2010–2011, it can be observed how over the total, a percentage of 45 % can be attributed to liquid fuels deriving from oil, 32 % to natural gas consumption and 22 % to the consumption of electrical energy.

From this short analysis, it can be pointed out that the Sicilian industry is strongly characterized by "Energy Intensive" industries (Refineries, Petrol-Chemical, Cement, etc.), which are not likely, for their production needs, to reduce their consumption. For these sectors a deeper insight is needed to understand what might be the margin of improvement. It is thus clear that to achieve the targets set out in the "Burden Sharing Decree" (Italian Ministerial Decree March 15th 2012), it is necessary to operate in other sectors which make up about 65 % of the entire energy consumption in the region.

---

[2] The Joint Research Centre is within the Covenant of Mayors the european institute for the validation of the SEAP that are elaborated and presented by the municipalities.

Specifically, this consumptions are related to transport, 40 %; residential, 23 % and tertiary, 37 %.

The analysis of energy certificates for buildings, which certifies the energy performance of the buildings in the region, up to 2012 shows that 86.9 % are G-class buildings, while only 0.5 % are in class A.

The majority of the building stock of Sicily is, in fact, composed of post-war buildings built during the economic boom (1950–1970 approximately) and generally consists of buildings with reinforced concrete frame construction and wall plug in blocks of tuff stone or perforeted bricks, walls without insulation, single-glazed windows and no technical measure to limit solar gains, thereby resulting in buildings with very low energy performance.

The energy consumption often connected to the heat dispersion from the walls and to the absence of heating and cooling systems, in most cases, is increased due to the use of electrical heating and cooling systems that are largely less efficient as compared to the high efficiency systems currently available on the market.

As far as renewable energy sources systems in residential buildings in Sicily are concerned, even if Sicily shows a favourable level of solar radiation (average yearly solar radiation on the horizontal plan 1.724 kWh/mq), the solar thermal systems for the production of sanitary hot water for residential use are installed only in 0.3 % of the total users in the Region. More frequently in Sicily electrical energy and gas are used for heating and small PV systems aren't much employed for the production of electrical energy.

The Sicilian Region has various projects that aim at sustainable and smart planning in different areas (Riva Sanseverino et al. 2014), including the Factor 20 Project.

The project, which involves the collaboration of three Italian regions (Lombardia, Basilicata and Sicily) is realized with the contribution of the European program *LIFE+ Environment Policy and Governance*, a financial instrument set up by the European Commission for the development, implementation and updating of the environmental policy of the European Union.

Factor 20 is aimed at defining a set of operational tools to support the planning of regional and national policies for the reduction of greenhouse gas emissions, the reduction of energy consumption and dissemination of renewable energy sources.

The project aims at promoting an integrated approach to the planning and monitoring of the results for the achievement of sustainable energy objectives set by the European Union by 2020, with reference to the well-known "20-20-20 climate package", and in particular to the recent 2012/27/UE Directive on Energy Efficiency outlining the goals for energy requalification in buildings.

The proposed approach involves different territorial decision levels in order to achieve common goals.

The Factor 20 Project aims at the identification of local actions that can effectively struggle against climate change (Factor 20 Project 2013).

The project, running between January 2010 and January 2014, is also dealing with:

- Involvement of local authorities in the experimental phase promoting the adoption of Local Action Plans that should identify an adequate set of concrete actions, coherent with the priority measures promoted at regional level and aiming at reaching the 2020 objectives, guaranteeing their measurability and continuous monitoring
- The action is implemented by means of the implementation of a Dedicated Information System (named SIRENA Factor 20).
- Promotion of actions aiming at raising the consciousness of local stakeholders about the importance of the CO_2 factor as a strategic element and as a priority for the definition of policies and actions for sustainability.

The Sicilian Region has identified the priority actions and measures to be included in the Local Action Plans among which the action about the "Improvement the energy and environmental quality of buildings, through the adoption of standards in the Urban Building Regulations ensuring greater sustainability and energy efficiency of buildings" (Vaccaro 2013).

5.3 The Regulatory Frame in Italy in the Field of Energy Efficiency in Buildings and in the Use of Renewable Energy Sources Generation

The interest towards a sustainable change in the building sector is now remarked in the contents of the Directive 2010/31/UE "Energy performance in buildings" (EPBD recast, active since July 2010 replacing the Directive 2002/91/CE) and by the new and more recent Directive 2012/27/UE[3] about energy efficiency that will be soon transposed into the Italian national legislation.

In Italy, on June 6th 2013, the Decree n. 63 has come into force, coordinated with the conversion Law nr. 90, into force since August 3rd 2013, that is the national Italian transposition of the Directive 2010/31/UE.

The Decree applies to the public and private buildings and dictates rules about energy performance of new buildings, important restructurings and energetic requalification, updating the decree 192/2005 transposing the Directive 2002/91/CE (old EPBD) that till now, with various updates, is still the national regulatory benchmark.

---

[3] The Directive, inter alia, raises the need to "increase the rate of restructuring of real estate, as the existing building stock represents the individual sector with the greatest potential for energy savings" stating also an annual rate of 3 % of mandatory restructuring for buildings owned by their central government in order to improve their energy performance.

Among the novelties, new energy performance certificates (APE)[4] compliant with the directive's requirements are introduced, replacing the old ones (ACE). Moreover in the same Decree new national guidelines for energy certification of buildings are announced.

Up to the issuing of actuation decrees concerning the calculation of energy performance indicators, are still in force the existing methodologies for the calculations of energy performance of buildings (Decree 59/2009).[5]

Therefore, in this field, the regulatory framework in Italy is not completed.

The last Directives, indeed, are strongly pushing towards a strong change involving the entire industrial chain of the building sector. The definitions provided by the EPBD recast are really impressive:

- *nearly zero-energy building* means a building that has a very high energy performance. The nearly zero or very low amount of energy required should be covered to a very significant extent by energy from renewable sources, including energy from renewable sources produced on-site or nearby;
- *energy performance of a building* means the calculated or measured amount of energy needed to meet the energy demand associated with a typical use of the building, which includes, inter alia, energy used for heating, cooling, ventilation, hot water and lighting (Directive 2010/31/UE, named EPBD Recast).

As well as the constraints for each EU member state to elaborate a useful strategy for the transformation of the all public and private buildings by 2050 into more efficient buildings (Directive 2012/27/UE).

It is thus necessary to set out a new paradigm for the design of new buildings as well as the maintenance of the old ones. The new integrated approach mixes the classical guidelines about geo-localization and orientation with Life Cycle Assessment[6] analysis, use of eco-friendly materials but also active measures such as the installation of Building and Home Automation systems. The latter systems allow the smart management and continuous monitoring of the energy performance of buildings. All envisaged measures and design approaches must take care of the expected return on investments, thus taking care about economic sustainability.

---

[4] As stated by the new Directive, to learn about the energy performance of buildings no longer the technicians have to refer to 'energy efficiency', but their 'energy performance' (annual amount of primary energy actually consumed or expected to be required to meet with a standard use of the property, the different energy needs of the building, winter heating and summer, the preparation of hot water for sanitary use, ventilation, and for the tertiary sector, lighting. Such a quantity is expressed by one or more descriptors that also take into account the level of insulation of the building and the technical characteristics and technical systems.

[5] Referring to the technical norms UNI TS 11300 parts 1, 2, 3, 4, reccomendation CTI 14/2013 and technical norm UNI EN 15193.

[6] The Life Cycle Assessment is a methodology that evaluates a set of interactions that a product or service has with the environment, considering its entire life cycle that includes the stages of pre-production (and thus also the extraction and production of materials), manufacturing, distribution, use (and therefore reuse and maintenance), recycling and final disposal. The procedure LCA is internationally standardized by ISO 14040 and 14044 norms (http://it.wikipedia.org).

The role and the potential of home automation systems in implementing energy efficiency are universally recognized since they cover an area that may be strategic in reaching the objectives posed at European level about energy efficiency.

A recent analysis from Schneider Electric has evidenced that in Italy due to the limited efficiency of the power transmission and distribution system, 1 kWh of electrical energy saved in consumption, is equivalent to 3 kWh saved of generated electrical energy. Besides, the use of low energy light bulbs into a room where nobody stays is still a waste of energy that can be limited by using home automation systems. In Italy, in this regard, was recently developed a guide for evaluating the effectiveness of the building automation systems (Italian Technical Standard CEI 205-18 2011).

Indeed, the greatest energy savings can be obtained when active (building automation systems) and passive meausers (low energy light bulbs for example) are used contextually.

Considering that buildings consume about 25 % of the total energy used in various fields, it can be realized that in absolute terms the contribution of automation systems, can be decisive for Italy in the achievement of energy efficiency targets by 2020.

The historical building heritage of Italy indeed in most cases would not allow the utilization of invasive passive measures. In these cases, it would be advisable to implement only active measures. The latter if implemented in the building, can bring savings in thermal consumptions for heating and cooling up to 26 %.[7]

To do so, the government should put in place a system of incentives for these technologies, allowing to make them accessible both to public and the private users, so as to enable the market and making them cost-competitive.

There are many expectations in this regard, in Italy about the Italian transposition of the Directive 21/2012/UE, which could give them an organic position. The change, however, should also cover the new Energy Performance Certificate (EPC), which is mandatory in Italy since 2012 to be issued in the event of sale or rental of a property and in cases of major building renovations (insisting on more than 25 % of the surface building envelope), or for new buildings. The idea is to make the EPC the identity document of the economic value of a property.

The system currently ruling the consumption calculations for such a certificate (Italian Technical Standard UNI TS 11300-1 2008; Italian Technical Standard UNI TS 11300-2 2008), is based only on a "quasi-steady modelling" of the building-plant system and of surrounding environment, making, on the one hand, easier consumption metering and the energy comparison between different buildings, but on the other, ignoring almost completely the ability to recognize and make the best

---

[7] The Italian norm CEI 205-18, transposing the european norm EN 15232, is a practical guide to the use of Building Automation technologies in buildings, clearing out how starting from a building without automation measures for thermal and electrical plants (class D building according to the EN 15232), it is possible to attain large energy savings by means of the installation of automation functions. As an example the CEI norm states that it is possible to get a reduction of thermal energy consumption of 26 % by installing class A automation systems.

use of the real environmental conditions and use of buildings, which are by their nature variables of the problem.

This leads, then, to a design that uses standard solutions, compared with home automation solutions that allow the automatic adaptation to environmental variables and use of the buildings, thus contributing significantly to arrest the increase in efficiency and the evolution of a real estate market in which highly energy efficient technological systems can be valued like other valuable features of the buildings.

In what follows, a summary of the current European directives on energy efficiency in buildings and on the installation of renewable energy sources systems and of the relevant transposition norms in the Italian context is reported.

The directive 2002/91/CE "EPBD" was adopted in Italy with two legislative acts: the decree of August 19th 2005 nr. 192 then modified by the decree December 29th 2006 nr. 311, which has proposed the general framework for the implementation of the directive and has updated the efficiency requirements and the compulsory transmittance values for the parts of the building envelope. These decrees have then been integrated by two decrees: the decree may 30th 2008 nr. 115 (then modified by the decree 56/10) and the decree April 2nd 2009 nr. 59. The first implementing the directive 2006/32/CE and introducing novelties in terms of volume prizes in case of the use of systems to increase the efficiency of the building envelope; the second actuating some of the points of the decree 192/2005, by defining the technical norms to be used for the calculation of the energy performance of the building.

The decree nr. 63/2013 coordinated with the Law nr. 90/2013 transposes the EPBD Recast (Directive 2010/31/UE); they introduce fiscal detractions for energy efficiency interventions. They also claim the need of an Action Plan to make compulsory the Near Zero Energy Buildings construction in public administrations by 2018 and in the private sector by 2021. As far as the RES integration in buildings is concerned, the decree 28/2011 implementing the directive 2009/28/CE is still valid. In the decree at articles 11–12 are specified the obligations[8] when installing RES in new or strongly restructured buildings.[9]

---

[8] In the case of new buildings or buildings undergoing major renovation, facilities for thermal energy production must be designed so as to ensure the respect of the contemporary coverage, through the use of energy produced by plants fueled by renewable sources, of 50 % consumption required for domestic hot water and defined percentage of the sum of the estimated consumption for domestic hot water, heating and cooling, variables with respect to the date of presentation of the building to the municipality. These obligations can not be fulfilled by renewable energy plants that produce only electricity which, in turn, devices or equipment for the production of domestic hot water, space heating and cooling.

[9] Article 2, comma 1-m of the decree 28/2011: "building strongly restructured" is a buildings falling into the following categories:
(i) existing building with a floor area greater than 1,000 squared metres, subjected to integral restructuring of the building elements of the envelope; (ii) existing building subjeced to demolition and rebuilding also in extraordinary maintenance.

5.4 The Definition of Guidelines for the Sustainable Revision of the Municipal Building Regulations: The Case of Sicily

The compliance to the principle of building in sustainable way, namely the achievement of environmental quality in building and town planning should be implemented by means of a regulatory and organizational model for the design and verification of the energy performance of buildings.

The main tool in building production is the municipal building code, seen as an operational tool that rules the transformation of the urban fabric, ensuring the application of principles and criteria to be delivered at the municipality level.

The common practice for the update of the municipal building code is that to define a specific *annex* that specifies the measures required to get closer to the near zero energy buildings as defined by the European directives. The specifications should address the performance issues and the support measures, namely incentives, carried out by the municipal administration to back up specific energy efficiency improvement actions.

The definition of Guidelines for the revision of the municipal building codes of a whole Region in a member state of the EU is an efficient tool to reach a distributed and concrete change in the territory even if it intrinsically has the limit that it can only define generalized technical and operational contents, that must be adequate in the different territorial contexts of the region and thus need a further specification to make them really applicable to the local features of the territory where the relevant municipal building regulation will apply.

To achieve the required energy performance, the bioclimatic design principles seem the most appropriate choice.

Bioclimatic design takes into account local environmental features to maintain comfortable conditions inside the buildings. Such design technique, in every phase of the building process, relies on energy from renewable sources. In the same way, it deploys all the measures that minimize the energy consumptions and reduce the losses while optimizing the potential of the materials employed exploiting the local climate features.

In this aim the climatological analysis of the design site is essential.

At local level, the climatological features throughout the Sicilian region, as an example, can strongly vary; therefore the following considerations refer to a temperate climate, characterized by hot summers and mild winters.

On the Sicilian coasts (especially in the south–east), the climate is strongly affected by the African currents, which cause very hot summers.

Along the Tyrrhenian coast and in the internal areas, the temperature is lower, the winters are colder and the rain is more intense.

The municipal administration that wants to define in greater details the principles of bioclimatic design for building in a specific territory, considering the national regulatory framework, will have to execute an accurate study starting from the climatological knowledge of the area.

Through the analysis of all the possible relations between the building and the environment (such as the geographic position of the site, the climatic zone, the meteorological parameters, the morphology of the land, the presence and use of green areas, features of the urban layout, features of the local materials used for buildings, building techniques, and so on) the most adequate parameters of bio-climatic design will have to be specified.

The cited annex to the municipal building codes will have to comply with the other territorial planning tools such as the Municipal Energy Plan, if existing. Also, in urban contexts where historical heritage is huge, such as in most Italian cities, the annex will have to account for such context in order to preserve the historical and cultural identity of the site, giving value the constructive elements of the historical buildings or in the traditional features of the fabric, that typically find full compliance with the principles of bioclimatic architecture.

The Administration to make the new sustainable building code truly operational, will have to implement control and verification procedures of the projects aiming at verifying the correspondence between the contents of the building regulations and the project, both during the presentation as well as in the realization phase. This is a crucial step in order to set out a real change in the way the buildings are created and managed.

To make the procedures easier the Administration can set out a synthetic *checklist* with the mandatory interventions in terms of energy efficiency included in the annex, the checklist will be completed and presented by the expert as annex to the technical report of the project.

The checklist will contain the climate, geometrical, energy needs data and also the features of the building envelope and of the technical systems. The technician from the Administration on its side will have at a glance all the data to check the correctness of the procedure both during the preliminary presentation phase of the design and in the realization phase in the building yard.

5.4.1 The Sicilian Context

In the aim of creating the basic knowledge and make applicable at territorial level a sustainable revision of the municipal building regulations the work has been carried out in different phases. Starting from the analysis of the municipal building regulations of some local administrations in Sicily, the most critical aspects for the development of energy efficiency related measures and implementation of renewable energy sources have been searched. The executed analysis has evidenced that there are no real regulatory obstacles for such measures, the main problems being the inertia of local administrations towards any change and the consequent missing transposition of the national and European legislation.

If on one side, indeed, the legal framework of energy efficiency is continuously changing, with European directives and national transpositions that make it difficult for local administrations the continuous update, on the other, the case study of

Sicily is particularly critical, because the existing building codes are not updated to the new quality control system in building production and thus they only refer to volumes, surface dimensions and heights (Vaccaro 2013).

Based on the data about the EPC available from the regional department for Energy (year 2012) and already discussed, more than 80 % of buildings show a very low energy performance (G class[10]). Such indicator underlines how in the Sicilian region the respect of the minimum energy requirements in buildings is still unusual. Currently in Sicily the market and the production chain connected to energy requalification of buildings is still not developed, there are no professional skills in this area and the high education does not provide tools to update the knowledge in the field of the future technicians and professionals.

Such contextual situation is more relevant in the small centres of the region, where often the local building experience is strongly consolidated.

Going at local level, the main critical aspect is the missing knowledge and understanding about the technical aspects supporting the energy efficiency issues.

Professionals that are not able to welcome innovation often manage the technical sections of administrations and the local stakeholders consequently are not encouraged to update the way of working in the building sector.

The considerations above expressed are of basic importance to understand why in Sicily, more than in other areas, it is essential a coordination carried out from the regional department of Energy through suitable guidelines that can address the annex of the municipal building code of each urban centre.

The aim is to give guidance to local administrations that want to revise their Municipal Building Regulation in the light of sustainability and of the current legal framework and obligations towards Europe.

The latter being the essential starting point for local incentive policies of municipalities aiming at "near zero energy building" design.

Even if the *annex* to the municipal building regulation should only contain improvement measures, as compared to the existing limits fixed by the current legal framework, in the Sicilian context, the start of a transformation process must first of all enable *sustainable building*, gradually producing a change within the entire building production chain.

---

[10] The energy classification of buildings in Italy, till now, was executed referring to the quantity of primary energy per m^2, required for heating and sanitary hot water production. The energy scale for buildings ranging from A+ to G represents the index of global primary energy (EPgl) given by the summation of the indices of energy performance for winter heating (EPi)—representing the energy consumed in one year to heat one m^2—and for hot water production (EPacs). The latter value depends, largely, for the production means, namely if it is employed a centralized source or a local autonomous source, with gas supplied boiler or electrical boiler. It must then be considered that, the average energy consumption of the buildings realized before 1977 in Italy it is $200 \div 250$ kWh/m^2 per year (Class G). Such consumption must be compared with the average energy consumption in Germany, Austria, Switzerland and Denmark of $20 \div 50$ kWh/mq per year (in http://energaiacn.it/home/61-certificazione-energetica-e-valore-edifici.html).

Fig. 5.1 The regulatory framework for the municipal building codes and the energy annex

A constant information activities and awareness among local stakeholders are the enabling actions for implementing the above-cited changes by the administration. One of the main operational tools is the municipal building code that can be considered as a support tool for the design choices and performance verification and that can be better suited to the local situation.

The analysis of the current regulatory framework concerning the energy efficiency and the renewable energy production at European, at national and at regional level in Italy is the first phase for the definition of the municipal building code, see Fig. 5.1.

In this way it is possible to clear out what are the constraining boundaries that are currently ruling the energy efficiency in the building sector and how wide is the leeway for the local administrations, for setting out local regulations aiming at improving energy efficiency in buildings.

The *Regional guidelines* that will be described in this chapter aim at clearing out the cited boundaries in order to limit the free interpretation of the current regulatory framework and at the same time provide the basic structure of the *energy annex* to the municipal building regulation that can be implemented by each municipal administration as regulatory and addressing tool for the development of the urban context.

To support the definition of some performance improvement actions referred to the building envelope, some sectorial studies referring to the Passivhaus Standard in southern Europe about the limitation of energy consumption due to summer cooling, were taken as reference.

The proposed document, accounting for the existing regulatory framework and clearing out the European and national objectives on the issue of energy efficiency, will indicate the possible support measures that the administration can set out, will focus on the main critical aspects of the building sector in Euromediterranean countries and will give an indication for the issue of the energy annex to the municipal building regulation.

It must be here specified that, since Sicily is non homogeneous in terms of climatic zones (A, B, C, D, E) and construction features of the buildings in urban contexts, the Guidelines will propose some gross *numerical limits* leaving to the local administrations the specifications according to the specific climatic zone or construction feature or other local needs.

5.4.2 The Analysis of the Regulatory Framework in Energy Efficiency and Energy Production from Renewables Sources in Sicily

In the aim of identifying the constraining elements for the Guidelines it is required to carry out the analysis of the regulatory framework on the topic of energy efficiency in buildings and renewable energy sources installations in buildings.

In Sicily, at regional level, the regulatory framework is totally compliant with the national framework.

The regional law n. 6, 23/03/2010 "Norms supporting the building activity and building heritage rehabilitation" (the so called *Piano Casa*[11]) was the first and only regional norm dealing with the introduction in the building practice of sustainability. Its applicability nonetheless is limited to few municipalities showing special features. These are indeed located in large territories and have many inhabitants, in such cases the political inertia is won by external stimuli: citizens, professionals and business with knowledge and sensitivity to energy saving, environmental sustainability and the related benefits.

Through the decree July 7th 2010, the region has issued a document for the definition of the technical and building features for the interventions in green building, enabling tool for the actions of the *Piano Casa*. Such indications refer to the technical interventions to be used in case of demolitions and reconstructions of the buildings, and if they are implemented according to the green buildings techniques, it is prized by the local administrations by allowing a volumetric increase of the building from 25 % up to 35 %. In the latter case, the building must also install renewable energy systems allowing the complete coverage of energy needs during a standard use of the building.

The main drawback of such document, that has probably limited its application, is that it provides a set of actions concerning the energy issue, water, waste, health and comfort, without giving technical specifications that can actually address professionals thus allowing the municipal administration a suitable control over the projects proposed by the privates. Moreover the cited document does not clear out what are the minimum requirements that a building must hold to be defined built in a "sustainable way". If, such as it happens in Sicily, the regional law transposes the national law, without giving further indications about the local context, the document containing the guidelines must compensate this gap in order to meet the European and national requirements in the specific regional context.

---

[11] National law, whose first version was issued in 2009 (Berlusconi government) in order to support the building sector offering the citizens the possibility to restore or enlarge the volume of buildings with a strong simplification of burocracy, overcoming the local municipal urban plans and introducing energy efficiency measures as a couterpart. Each italian region has then issued a regional law, concerning the matter. In Sicily the possibility to benefit from such law is extended till August 2014.

5.4.3 Prizes and Incentives for Energy Efficiency in Municipal Building Codes

It is well known that the incentives are the most important tool for the development and consolidation of the innovation technologies market; the case of photovoltaic panels market development in Italy following the incentives "Conto Energia" in Italy have brought up to now a strong reduction of prices and the widespread of the technology over the national territory (including the Sicilian region), even if up to now the so called grid parity[12] has not yet been reached.

Incentives and prizes within the municipal building code are necessary to support the widespread of energy efficiency measures that differently would have too long return on investment times especially for privates.

Since public administrations are going through a major economic crisis in all Europe and most critically in Euro Mediterranean regions and considered that any kind of prize or incentive is currently turned into a reduced income for local public administrations, it is fundamental an economic evaluation of the measures. The latter must indeed be chosen so as to immediately produce a quantifiable return on the investment in energy saving terms, in environmental benefit terms while giving rise to a sustainable development of the territory.

The incentive must account for the applicability of the technologies on the base of the existing urban planning tools, of the prevailing building typology as well as the possible acceptation from the community of a certain type of intervention. Indeed, for many of the actions that contribute to reaching the environmental sustainability, it is fundamental the degree of involvement of the citizens, that on one hand must be motivated by the economic interest, on the other must be informed about the use of the technologies.

An analysis carried out by the manager of electrical services in Italy (Gestore dei Servizi Energetici, GSE) not considering the national incentives, compares the energy efficiency measures by means of the economic return for the investor versus benefits for the community for avoided environmental costs.

The analysis shows, for example, that the substitution of shutters has too long return on investment times, setting almost to zero the return for the investor. Assuming the technical lifetime of the intervention of 30 years and the cost of the investment per unit of glass shutters of around 400 €/mq, it can be easily seen that, even with the current incentive of reduction of taxation of an amount equal to 65 % of the entire cost supported, considering an actualization rate of 5 %, the return time of the investment is of 23 years. Such value is largely higher as

---

[12] Grid parity occurs when an alternative energy source can generate electricity at a levelized cost that is less than or equal to the price of purchasing power from the electricity grid. The term is most commonly used when discussing renewable energy sources, notably solar power and wind power. Reaching grid parity is considered to be the point at which an energy source becomes a contender for widespread development without subsidies or government support. It is widely believed that a wholesale shift in generation to these forms of energy will take place when they reach grid parity (http://it.wikipedia.org/wiki/Grid_parity).

compared to the 15 years considered to be satisfactory in the building sector to consider the implemented measure economically profitable.

The incentive tool, moreover, allows to clear out the political orientation and the territorial development direction that a given local context want to pursue, producing as side effect the economic development in the area and costs reduction, due to the widespread of the supporting technologies.

From the analysis of the Italian Report ONRE[13] 2013 (Cresme Ricerche s.p.a e Legambiente 2013) related to energy innovations in buildings, comes into light an interesting definition of those that are considered the current incentive measures used in the building sector within the building regulations. The prizes or incentives can be classified into three main categories.

The first is related to economic incentives, as reduction of the secondary urbanization burdens[14] or elimination of the building contribution, most often increasingly recalibrated depending on the level of attainable energy saving, of eco-compatibility of materials and of the adopted constructive technologies, as well as of the buildings features surpassing the mandatory limits set by the regulatory framework.

All these are used as incentive measure in 45 % of the new municipal building codes.

The second category concerns incentives such as urban planning, volumetric prizes, for which recognizing the improvements in the energy performance of the building is granted an extension that will not be calculated as floor area. The deduction from the whole volume, of the increased volume required by the installation of new plants or the creation of thermal coating or of other energy efficiency measures, in derogation from the existing general urban plan is also a very common measure in the new municipal building codes in Italy.

The third class of incentives is the direct financing through calls for some given measures; it is the case of calls for direct contributions.

Some municipalities, such as Bari in southern Italy, has set incentives for those implementing sustainability measures in building such as the reduction of the waste taxes.

It is also necessary to recall, in this context, the current national situation concerning building restructuring and energy efficiency measures.

The public incentives framework is up to now very favourable, symptom that these measures are considered relevant within the national energy policies and kick-starting economy measures in Italy.

Just as an example, the taxation reduction on building restoration (36 % of invested capital), on energy efficient restoration of buildings (65 % of invested capital) and the Conto termico (supporting small measures on existing buildings, to increase energy efficiency and the production of thermal energy from renewable energy sources).

---

[13] National Observatory of Building Regulations.

[14] Concerning the implementation of social services supporting a urban site, such as schools, churches, or green areas.

As already mentioned, Sicily has also still active the Piano Casa, providing volumetric prizes for demolition and rebuilding according to green building techniques.

5.4.4 A Benchmark for the Elaboration of Performance Standards to Improve the Current Regulatory Framework

From the analysis of the energy consumption in buildings of the Sicilian Region, not surprising that the peak comes during the summer season in the hours of major irradiation. Such data can be essentially connected to the climate features of Sicily that comprises different climatic zones, but is characterized by very hot summers all over the region.

Considering that in the national regulatory framework in Italy the energy efficiency techniques are related to consumptions and comfort requirements during the winter season, due to two different reasons. On one hand, the knowledge about thermal modelling of buildings concerning hot climates is more difficult, on the other, the European countries that are more attentive to environmental issues and energy efficiency are in northern Europe.

Therefore, the measures to improve energy efficiency of buildings to be included in the regional guidelines in Sicily have been defined based on a scientific study elaborated by the eERG group of the polytechnic university of Milan (Italy) within the project Passive-On.[15]

The reference document is entitled *Passivhaus for southern Europe –Guidelines for the design* (end use Efficiency Research Group, eERG 2007) has revealed that it is generally possible to limit without too much difficulty the thermal loads in Southern Europe countries bringing the demand of net useful energy demand for heating and cooling to 15 kWh/m^2 year, while the demand of primary total energy (rooms heating, domestic electrical appliances, lighting and sanitary hot water) is equal to 120 kWh/m^2 year. All this to keep a summer operational temperature equal to 20 °C in winter and below 26 °C in summer.

What results from the analysis is that the cooling thermal loads can often be covered with only passive measures (external insulation, solar screens, evaporative cooling, etc.).

Even if the Italian *Passivhaus* inherits many of the concepts of the German *Passivhaus*, the study has evidenced the need to modify some detail specifications. In general the Italian mild climate allows reaching the comfort and the energy limits of the *Passivhaus* standard using less constraining for what concerns:

- The insulation of opaque surfaces: while a typical German Passivhaus requires 25–35 cm of insulation external coating and 30–40 cm on the roof, in Milan (latitude:45°28′38″28 N; heating degree days: 2404; cooling degree days: 482;

---

[15] Research and dissemination project within the SAVE Intelligent Energies program, developed between 2005 and 2007, aiming at the promotion of passive houses in hot climates proposing to analyze the way to extend the passivhaus design concept to Southern Europe.

Table 5.1 Results of the *Passivhaus* concept translated to the Mediterranean area (end use Efficiency Research Group, eERG 2007)

Location	Air permeability of building (n_{50}) (h^{-1})	Transmittance of the building envelope (U-value) ($W/m^2 K$)				Insulation levels (cm)		
		Roof	Walls	Basement	Glass	Roof	Walls	Basement
Milan	1	0.134	0.135	0.134	1.400	25	25	25
Rome	1	0.200	0.300	1.000	1.400	16	10	1
Palermo	1	0.540	0.420	1.340	1.400	5	6	0

average winter temperature: 2.8 °C; average summer temperature: 21.7 °C) it is possible to satisfy the standard with less thick coatings both on external walls, basement and on roofs (25 cm). In Palermo (the capital of Sicily, latitude: 38°6′43″56 N; heating degree days: 751; cooling degree days: 642; average winter temperature: 13.9 °C; average summer temperature: 25.5 °C) such thickness can be reduced up to 5–6 cm, if the mechanical ventilation with heat recovery is kept, or alternatively the mechanical ventilation can be eliminated using a thicker coating (about 25 cm). In what follows the Table 5.1 below summarizes the results of the *Passivhaus* concept translated to the Mediterranean area.

The study concluded in 2006 shows values of external insulation and air permeability required to achieve the *Passivhaus Standard* in Milan, Rome and Palermo, using a thermal insulation with conductivity 0.037 W/mK and mechanical ventilation with heat recovery.

- The airtightness of building[16]: the standard and best practice of central Europe require that the building coatings limit the air exchange to a maximum of 0.6 volumes per hour for a pressure difference of 50 Pa (n50 < 0.6 h^{-1}). But the limit for Milan and Rome of n50 equal to 1 h^{-1} (referring to optimal tightness) is even too conservative in southern cities such as for example Palermo since the winter external design temperature overcomes 0 °C (according to the norm UNI 5364 in Palermo it is equal to 5 °C) and as it is well known the air loss, namely an uncontrolled air flow through the walls, doors, the windows and the

---

[16] Measured with the airtightness test (Blower Door Test) allowing to define the equivalent air flow for infiltrations at a pressure difference of 50 Pa (V_{50}).The number of air turnovers per hour at the pressure of 50 Pa is indicated with the index n_{50}. Such values allow to evaluate the quality of the building coating under the airtightness profile. The lower the value the more efficient the coating. Indeed a cold air infiltration from outside or the leakage of hit air from the inside, due to not perfect sealing, become a concentrated flow of water and steam. As a result the condensation inside different components such as walls, node wall-subframe, node subframe shutters, thermal insulation, roofs, etc.). The effects are quite negative: the thermal conductance of a material increases with the internal increased humidity and the retention of condensed water in the structures damages the materials, favours the upcome of mold, causes a decay of the living comfort. Moreover molding phenomena in wooden structures can take place or the fast deterioration of insulating materials (Ronchini 2012).

roof, is a function of the pressure difference and thus of temperature difference between internal and external space.

- The thermal transmittance of the transparent surfaces: the triple glasses normally employed in centre Europe can be replaced with double glasses with low emissions, or even better to reduce the too large solar gains in summer without penalizing the sunlight it is advisable to use solar control or selective glasses reflecting the close infrared radiation (namely the thermal part of the irradiation) but still keeping the transparency to the light, namely the visible part of the solar spectrum.[17]

5.4.5 The Structure of the Guidelines for the Definition of the Energy Annex to the Municipal Building Codes in Sicily

Based on what was said before, it is shown below the structure of the guidelines document.

The document will comprise 4 areas:

- **Area 1. Environmental sustainability and context appraisal.**
- **Area 2. Energy performance of building envelope.**
- **Area 3. Energy performance of technical systems.**
- **Area 4. Renewable energy systems.**

Each area is composed of articles; each article contains voluntary and mandatory features (both can be distinguished by the Administration through a specific reference into the text of the document). In this aim the reference for the technical choices, as already mentioned, was that of the Passivhaus Standard for southern Europe in the aim to focus the problem of summer cooling and of the associated consumptions also including improvements having as final aim the consumptions reduction in buildings during summer time.

The mandatory requisites imply the respect of the regulatory framework and constitute the complex of the mandatory norms to which each project must comply. These are subjected to continuous update each time variations on the reference legislation come up at regional and national level.

The municipalities, based on the adopted local development policies and on the local resources and the municipal balance, can insert, within its own municipal building code the proposed indications (indicated in italics in the examples shown in the following illustrative articles). In this way it will be possible to define performance levels improving the energy efficiency in order to support designing and realizing green buildings with high energy performance. The support can tale place through prizes and incentives as discussed before.

Another basic element is the increase of the level of surveillance and the application of fines, which must be carried out by the municipal administration.

---

[17] The degree of selectivity is described by the ratio between visible transmittance and solar factor.

It would therefore be necessary the continuous education of technicians, at regional or municipal level, aiming at the education in the field of public and private operators, so as to increase the speed of change and the correct application of the proposed measures. In this sense the municipal building regulation else than aiming to discipline and disseminate, also helps to improve the knowledge in terms of "sustainable building" among designers and builders. Therefore, at the same time, the definition of the energy annex of the municipal building code, each administration will have to define the formal procedures to be followed by the interested parties and the necessary contents for the release for certifications and the verification of the requirements.

In what follows, it is outlined a more detailed structure of the document with the indication of the articles.

Area 1. Environmental sustainability and context appraisal

Art. 1 Building orientation

Art. 2 Natural lighting

Art. 3 External microclimate control

Art. 4 Usage of arboreal types

Art. 5 Water saving

Art. 6 Usage of eco-friendly and recycled materials

Area 2. Energy performance of building envelope

Art. 7 Verification of energy performance in design phase

Art. 8 Ratio dispersion surface—volume at controlled temperature

Art. 9 Thermal insulation of buildings (building features of the envelope—winter)

Art. 10 Thermal insulation of buildings (building features of the envelope—summer)

Art. 11 Shutters performances and shadings control

Art. 12 Roofs performances

Art. 13 Use of passive cooling measures

Area 3. Energy performance of technical systems

Art. 14 High performance heat generation systems, thermal centralized systems and metering systems

Art. 15 Thermal regulation systems

Art. 16 Low temperature systems

Art. 17 Mechanical ventilation systems

Area 4. Renewable energy systems

Art. 18 Solar thermal systems

Art. 19 Renewable energy sources for electrical energy generation

Art. 20 Electrical infrastructures for electric vehicles recharging.

In what follows, are reported, as an illustrative example, some of the articles of the Guidelines proposed for the Sicilian Region.

Illustrative example of contents of the articles within **Area 1. Environmental sustainability and context appraisal.**

Art. 3 External microclimate control

1. *The built environment influences the climatological features of the site, based on the urban layout (as a function of the position and the density of the neighbouring*

buildings related to shadings, winds and local temperatures). An extended area of built environment can, indeed, generate an increase of local temperatures (urban heat island effect). In this aim it is advisable to maximize the open spaces.

2. *For new buildings and for those subjected to demolition and total reconstruction it is strongly suggested to create an open space with a surface of at least 20 % of the design area creating shades by suitably planting trees. The phenomenon is also limited by green roofs or employing reflecting materials for roofs. Also the insertion of walking routes favouring the natural ventilation giving rise to cooling of external walls of the buildings. It is also suggested, to improve the microclimate around the buildings, the use of cool surface materials for external floor surfaces, among which: grass fields, reinforced grass paving system, bricks, clear stones, cobblestones, such as also around the built land, especially on the surfaces exposed to summer solar radiation from 12 a.m. to 4 p.m. (standard time). The latter measure, using higher reflectivity materials for such external surfaces, mitigates summer temperature peaks with a reduced absorption of the solar radiation in the infrared spectrum.*

These indications refer to the bio-climatic architecture, defining design criteria for the control of external microclimate control close to the building.

The above measures produce a natural cooling effect during summer. The plantings will be located also to protect the walls from winds during winter and the direct solar radiation during summer.

They will be chosen so as to direct the summer winds towards the building and to favour the natural ventilation inside the buildings (such notations are specified in article 4 of the proposed guidelines). As it can be noted, the above rules are in italics to indicate that they do not include any mandatory clause deriving from the current national legislative framework. Therefore the local administrations can decide whether such rules are to be set as "voluntary" or "mandatory" within the annex to the new municipal building code. The Administration can also devise some incentive measure to support the implementation of voluntary rules.

Illustrative example of contents of the articles within **Area 2. Energy performance of building envelope.**

Art. 9 Thermal insulation of buildings (building features of the envelope—winter)

1. For all classes of buildings if newly built or strongly restructured or under extraordinary maintenance of the coating[18] (as an illustrative example, the

---

[18] The interventions are described in article 3, paragraph 2, letters (a), (b) and (c) number (1) of the Italian Legislative Decree n. 192/05 and modifications: (a) number (1) integral restructuring of the building elements constituting the coating of existing buildings with floor area larger than 1,000 m^2; (a) number (2) demolition and reconstruction in extraordinary maintenance of existing buildings with floor area larger than 1,000 m^2; (b) volume increase (gross heated) larger than 20 % of the entire existing building (the application is only related to volume increase); (c), number (1), interventions on existing buildongs such as total or partial restructurings and extraordinary maintenance of the building coating and volumetric increase out of what already considered in letters (a) and (b) (only for the part of coating or of volume object of the intervention) [Update of the Italian Legislative Decree n. 311/06].

Table 5.2 Limit thermal transmittance values in some building components, from the Italian Legislative Decree n. 311/06 (2006)

Climatic zone	Vertical opaque structures (U_{limit} W/m^2 K)	Horizontal and inclined structures (U_{limit} W/m^2 K)		Openable windows including frames (U_{limit} W/m^2 K)	Glass (Ug_{limit} W/m^2 K)
		Roof	Floors to unheated rooms or to the outside		
A	0.62	0.38	0.65	4.6	3.7
B	0.48	0.38	0.49	3	2.7
C	0.40	0.38	0.42	2.6	2.1
D	0.36	0.32	0.36	2.4	1.9
E	0.34	0.30	0.33	2.2	1.7
F	0.33	0.29	0.32	2	1.3

repointing of external walls, of external plasters, of the roof and of the roofs sealing), the opaque vertical structures bounding the coating must show values of thermal transmittance (U) in accordance with the Table 5.2 (the values refer to the current legislative framework in Italy), as a function of the climatic area.

Moreover for all categories of buildings except those for industrial, handicraft and similar uses, for the same above cited cases, for the horizontal and inclined opaque structures, openable shutters and similar elements and glasses bounding the envelope, the values of thermal transmittance (U) reported in the following table must be respected.

Such rules are mandatory except than for those buildings having social features: historical buildings, buildings for industrial and tertiary use when the rooms are heated for needs connected to the productive process or using the waste energy deriving from the productive process and that would not be used elsewhere, isolated buildings with floor area below 50 m^2.

The values of U must be respected with thermal bridge solved,[19] when the design of the coating does not require the correction of thermal bridging, the limit

---

[19] One of the key aspect of making low-energy buildings is the care given to the thermal insulation. That means using the right materials with the right thickness to provide the right thermal coating. But this is not enough. The thermal insulation must not have holes where heat would flow thus defeating the purpose of the thermal insulation. Fighting those thermal bridges is essential in making excellent quality low-energy consumption buildings. Thermal bridges are localized regions in a building which display increased thermal losses. They can be caused by components whose geometry, such as balconies, or whose materials, such as aluminium window frames without thermal break, have higher thermal conductivity. Thermal bridges are most often created by the structure of the building, at the junction of walls and floors, at the junction of walls and roof, in the corners or around windows if they are not properly installed. Interior thermal insulation is well known to create many thermal bridges that could be completely avoided for example by doing exterior thermal insulation (http://beodom.com/en/education/entries/fighting-thermal-bridges-or-how-to-make-better-buildings).

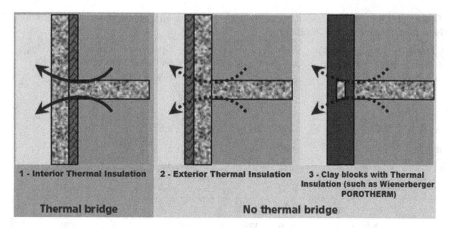

| 1 - Interior Thermal Insulation | 2 - Exterior Thermal Insulation | 3 - Clay blocks with Thermal Insulation (such as Wienerberger POROTHERM) |

Thermal bridge   **No thermal bridge**

Fig. 5.2 Example of thermal bridge and no thermal bridge in vertical opaque external structures (Reproduced from Beodom.com)

values must be considered as bounding the average thermal transmittance of the wall plus the thermal bridging. In Fig. 5.2 an example of a solved thermal bridge is given.

For vertical opaque external structures where limited areas must show a reduced thickness, under-windows or other components, the limit must be respected with reference to the entire calculation surface. For horizontal structures on the floor the values of U to be compared with the limit have been evaluated with reference to the system structure-ground. The automated walkable entries are excluded and they must be considered only to evaluate the changes of air.

It must be noted that if the edifice is built out respecting the limits imposed by the current legislation (Table 5.2) acceptable levels of energy performance could be reached. But a possible improvement of the transmittance of such component would allow reducing the heat winter losses thanks to a highly sealed envelope and to the elimination of thermal bridges. An increased sealing of the envelope on the other hand reduces the air changes of the building envelope, therefore, it may be required an additional ventilation system (such as for example a forced ventilation system with heat recovery from exhaust air).

The energy required for winter heating is also reduced if the design includes the possibility to exploit the solar heat gains through the glass shutters or glass walls installed on the south oriented walls, in this way heat losses may also be reduced if the north oriented walls with glass have a more limited surface.

It is also important to underline the energy needs of the building due to summer cooling, are reduced if the solar earnings of the external surfaces are minimized thanks, as an example, to a highly sealed envelope (increasing its thermal capacity[20]) and to the sun shadings or by suitable trees plantings (notations and numerical parameters for design are reported in article 10 of the guidelines).

The thermal comfort during summer is maximized if the daily solar heat gains and the internal heat load collected during nigh time inside the buildings are extracted, using a nightly strategy of natural ventilation (through the windows opening) or forced (through mechanical ventilation systems).

A well-sealed and heavy structure is the ideal situation for the exploitation of the summer night cooling of the thermal mass of the building. During night time the air circulates through the building, using winds or natural air density gradients, or using a mechanical ventilation system (notations at articles 10 and 13), setting partly free the building from the heat held by its mass. The greater or smaller efficiency of the system must however be verified with reference to the local climatic features of the site where the Municipality resides.

From this short analysis, the sealing features of the building influence the building energy consumptions both for cooling and for heating; it is thus however necessary to evaluate the efficiency of a possible reduction of the value of thermal transmittance of the opaque components of the building considered as a whole.

In climates such as the one where the municipality of Palermo resides, as an example:

- A possible hyper-insulation of the building components (is a mechanical ventilation system with heat recovery is employed) does not improve substantially the winter behaviour of the building, since winter shows mild temperatures (on average the temperature is 13.9 °C).
- However an hyper-insulation would penalize the energy performance during summer, since as the temperature and the internal heating loads rise, such sealing would not allow the heat to be drained by the ground at a lower temperature. The analysis of the Passivhaus study in the Mediterranean area, shows how the choice of not insulating the basement is probably the most desirable in the yearly energy balance.
- A good insulation of the perimeter walls and of the roof is thus needed since it provides in some cases energy reductions going beyond 50 % as compared to the not isolated situation.
- The natural ventilation is less efficient due to the limited daily temperature fluctuation in Palermo (on average only 3 °C during summer) and it is thus necessary some kind of active cooling system to guarantee summer comfort conditions.

---

[20] The *thermal capacity* is the property that certain materials have to absorb and retain heat in time, and this is measured in relation to the number of hours of temporal delay between the input of the heat flux incident on the exposed face (external) and its release on the opposite side (inside). To be effective, the material mass must be able to ensure a delay in the passage of the thermal wave such that the heat is released inside during the coolest hours of the day. The delay of the thermal wave, due to the heat capacity, is called *displacement*, while the reduction of the temperature on the inner surface, with respect to the external surface temperature, is called *attenuation factor*.

Based on these considerations and being the use of mechanical ventilation systems with heat recovery not usually installed in common residential houses, an improved energy performance of the building can be reached employing thermal transmittance values that are lesser than those indicated in the Italian Legislative Decree n. 311/2006.

The measures of energy efficiency in the building sector can hardly be standardized since they are strongly connected to elements such as the type of envelope, the orientation and the technical plants features. In the aim of identifying an optimal situation for comfort, both during winter and summer, it is required to simulate the proposed measures through suitable software tools set able to reproduce the above outlined conditions, in order to identify the set of more profitable actions in terms of energy and costs savings.

Based on what was said before within the Guidelines the local administrations can find some indications about the improvement measures to be included into the energy annex of the new municipal building regulation. In this way, even if the indicated measures will not allow attaining Near Zero Energy Buildings, they will certainly contribute to reaching this goal for all buildings typologies.

The article 9 of the guidelines is thus completed, in its mandatory features reported in the table above and desumed from the current legislative framework, with some other paragraphs among which number 2, reported below. It refers to improvements actions that can be implemented and proposed by the municipal administrations:

2. *In all cases referred to in paragraph 1 (above), in order to ensure greater comfort conditions in both winter and summer and a consequent saving in terms of primary energy for heating and cooling in buildings, those actions referred to in paragraph 1 of this Article (art. 9 Thermal insulation of buildings) will be awarded if they provide for the vertical opaque structures, a thermal transmittance values (U) improvement from 40 to 70 % as compared to the limits set in the same paragraph, similarly to the roof.*

    The incentive will be as larger as larger is the percentage of improvement. The quantification of the improvement actions and the incentive policy must be defined when the new municipal building code is issued by the local administration.

As previously mentioned in Sect. 5.3 the implementation of BAC systems in buildings is one of the actions that can be implemented, in the design phase of new buildings or in the energetic requalification of existing buildings, in order to improve their energy performance.

In the following chapter, a study showing the effectiveness of the automation of technical systems in buildings of different energy efficiency classes is presented.

The results show that these systems can be included among the measures leading to greater energy efficiency in buildings (especially in buildings with low initial performance, which often the case of the buildings in the Sicilian Region). Therefore, such as the passive measures acting on the building envelope, they can be regarded as tools to improve energy efficiency and to be fully integrated into the municipal building code.

References

Cresme Ricerche s.p.a e Legambiente (2013) L'Innovazione Energetica In Edilizia—I regolamenti Edilizi comunali e lo scenario dell'innovazione energetica e ambientale in Italia, Rapporto ONRE 2013. www.legambiente.it/contenuti/dossier/rapporto-onre-2013-edilizia-sostenibile-crescita

Dall'O G, Galante A (2009) Efficienza energetica e rinnovabili nel Regolamento Edilizio Comunale. Edizioni Ambiente, Milano

End use Efficiency Research Group, eERG (2007) Passivhaus per il sud dell'Europa- Linee guida per la progettazione, Report Passive-On Project, Politechnical University of Milan (Italy)

Energy Report 2012—energy data in Sicily, Assessorato dell'Energia e dei Servizi di Pubblica Utilità- Dipartimento dell'energia—Regione Siciliana (2012). http://pti.regione.sicilia.it/

European Directive 2010/31/EU (2010) On the energy performance of building

European Directive 2012/27/EU (2012) On energy efficiency, amending Directives 2009/125/EC and 2010/30/EU and repealing Directives 2004/8/EC and 2006/32/EC

Factor 20 Project (2013). http://www.factor20.it/home

Italian Decree-Law n. 63 (2013) Urgent measures for the transposition of Directive 2010/31/EU of the European Parliament and of the Council of 19 May 2010 on the energy performance of buildings for the definition of infringement proceedings by the European Commission, as well as other provisions of cohesion social

Italian Legislative Decree n. 192/05 (2005) Attuazione della direttiva 2002/91/CE relativa al rendimento energetico nell'edilizia

Italian Legislative Decree n. 28/11 (2011) Attuazione della direttiva 2009/28/CE sulla promozione dell'uso dell'energia da fonti rinnovabili, recante modifica e successive abrogazione delle direttive 2001/77/CE e 2003/30/CE

Italian Legislative Decree n. 311/06 (2006) Disposizioni correttive ed integrative al decreto legislativo 19 agosto 2005, n. 192, recante attuazione della direttiva 2002/91/CE, relativa al rendimento energetico nell'edilizia

Italian Ministerial Decree March 15th 2012 (2012) Definizione degli obiettivi regionali in materia di fonti rinnovabili (c.d. Burden Sharing), Gazzetta Ufficiale 2 aprile 2012, n. 78

Italian Technical Standard CEI 205-18 (2011) Guide to building automation identification of functional block diagrams and estimation of related energy savings, 1st edn. CEI, Milano

Italian Technical Standard UNI TS 11300-1 (2008) Energy performance of buildings—part 1 Calculation of energy use for space heating and cooling, first ed., UNI, Milano

Italian Technical Standard UNI TS 11300-2 (2008) Energy performance of buildings—part 2 calculation of energy primary and energy performance for heating plant and domestic hot water production, 1st edn. UNI, Milano

Lumicisi A (2013) Il Patto dei Sindaci - le città come protagoniste della green economy. Edizioni Ambiente, Milano

Riva Sanseverino E, Riva Sanseverino R, Favuzza S, Vaccaro V (2014) Near zero energy islands in the Mediterranean: supporting policies and local obstacles. Energy Policy 66:592–602

Ronchini E (2012) Verificare la tenuta all'aria, Casa Clima magazine n. 35. http://www.casaeclima.com/ar_9022__RIVISTE-Rivista-CASACLIMA–CASACLIMA-N35.html

Vaccaro V (2013) Regolamento Edilizio. In: Rapporto energia 2013- monitoraggio sull'energia in Sicilia, Assessorato dell'Energia e dei Servizi di Pubblica Utilità, Dipartimento dell'energia, Regione Siciliana, pp 36–40

Economic Feasibility of Measures for Energy Efficiency

6

Eleonora Riva Sanseverino and Gaetano Zizzo

Abstract

In this chapter the economic impact of some measures for energy efficiency, both using automation of technical infrastructures (according to EN 15232) as well as implementing *passive* measures concerning the use of building materials and techniques (according to EN 15271), is studied. The chapter is organized as follows. First, a technical-economical study on the evaluation of the impact on residential buildings of Building Automation Control, BAC, and Technical Building Management, TBM, systems is presented. Then the same assessment of some passive measures that can be employed is carried out using the Passive House Standard for Mediterranean warm climates is shown. Numerical elaborations have been carried out for a sample house located in Palermo, Sicily (Southern Italy). The building is located in the EuroMediterranean area and thus all measures are referred to warm climates.

E. Riva Sanseverino (✉) · G. Zizzo
DEIM, University of Palermo, Palermo, Italy
e-mail: eleonora.rivasanseverino@unipa.it

G. Zizzo
e-mail: gaetano.zizzo@unipa.it

E. Riva Sanseverino et al. (eds.), *Smart Rules for Smart Cities*, Sxi 12,
DOI: 10.1007/978-3-319-06422-2_6,
© Springer International Publishing Switzerland 2014

6.1   The European Norms for Energy Certification and the *Passivhaus* Standard

Evaluating the effects of active energy efficiency measures, namely the automation of technical infrastructures of buildings, and those of passive energy efficiency measures, based on the use of particularly efficient building techniques and materials, is a quite challenging task. Besides, the evaluation here proposed refers to a residential house located in a Euro Mediterranean city (Palermo) and thus considers a warm climate scenario. The evaluations allow assessing the economic viability of the adopted strategy. The tools available for such numerical evaluations are set out for energy certification purposes and are referred to the European regulatory framework. As far as the active energy efficiency is concerned, a list of measures (automation functions) can be found in the technical standard EN 15232 (2012), while a list of passive measures in warm climates can be identified taking as reference the studies carried out for the *Passivhaus* standard in warm climates (Passive-On project 2013). A list of passive energy efficiency measures with the description of a methodology to numerically evaluate the effects can be found in the Deliberation of the Italian Authority of Electrical Energy and Gas (AEEG), EEN 09/2011. Such Deliberation implements the European framework on white certificates and energy efficiency. The work carried out in this chapter shows first how the control, monitoring and automation functions considered by EN 15232 can strongly influence the energy performance class, defined according to the standard EN 15217 (2007), of a single-family test house located in the Italian city of Palermo, overlooking the Mediterranean Sea. The analysis shows that the benefit that can be drawn from the installation of Building Automation Control, BAC, and Technical Building Management, TBM, systems depends on the type of existing technical appliances in the household and on the starting energy performance class of the house. The second part instead is devoted to the assessment of the effects of some of the passive measures identified taking as reference the studies on the *Passivhaus* standard in Euro Mediterranean warm climates, using the methodology outlined in the directive EEN 09/2011. The *Passivhaus* concept represents the highest energy standard today with the promise of reducing the heating energy need of buildings by around 90 %. Widespread implementation of the *Passivhaus* design would have a strong impact on energy saving.

6.2   Background on Energy Performance in Buildings: The Italian Case

The term "energy performance" indicates the amount of energy consumed to meet different needs associated with a standard utilization of a building (building heating/cooling, lighting, ventilation, etc.). The energy performance depends on constructive, thermal but also electrical factors like:

- the presence of electrical energy generation systems based on Renewable Energy Sources (RES);
- the electrical energy produced by Combined Heat and Power (CHP) systems;
- the presence of Building Automation Control Systems (BACS) and Technical Building Management (TBM) systems.

All EU Member States have defined methodologies for the calculation of the energy performance of buildings on the basis of a general framework provided by the European Performance in Building Directive (EPBD) 2010/31/EU.

It is important to stress, how the new EPBD recast, differently from the old Directive 2002/91/EC, puts greater importance on automation, control as well as monitoring systems. As a matter of fact, the new EPBD suggests the use of *these systems* as well as of *intelligent metering systems* for energy saving if a new edifice is built or of it undergoes major renovation in line with Directive 2009/72/EC.

Moreover, the already cited "Climate and energy package" enacted by the European Union (EU) in June 2009 sets a series of demanding climate and energy goals to be attained by 2020, also known as the "20-20-20" package (reduction in EU greenhouse gas emissions of at least 20 %, 20 % of renewable energy in EU final energy consumption, 20 % reduction in primary energy use to be achieved by improving Energy Efficiency).

A recently issued directive of this group is the Directive 2012/27/EU on energy efficiency. This Directive establishes a common framework of measures for the promotion of energy efficiency within the Union in order to ensure the achievement of the Union's 2020 20 % headline target on energy efficiency and to prepare for further energy efficiency improvements beyond that date. It lays down rules designed to remove barriers in the energy market and to overcome market failures that create an obstacle to efficiency in the supply and use of energy, it also provides for the definition of indicative national energy efficiency goals to be attained by 2020. The directive explicitly refers to automation as a tool to attain the cited objectives through the implementation of Demand Response[1] policies (Balijepalli and Pradhan 2011), while the wide spread application of smart meters and regulation systems is considered a cost-saving measure for energy gains and savings.

Following this direction, since 2007, the European Standard EN 15232 devises terminology, rules and methods for the estimation of the impact of BAC and TBM systems on energy performance and energy use in buildings.

The technical standard EN 15232, today in its second edition, gives a list of BAC and TBM systems functions that can affect the energy performance of buildings and defines two methodologies to evaluate the impact of these functions. Besides, the standard EN 15232 introduces four different efficiency classes for buildings according to BACS and TBM systems installation:

---

[1] According to the definition provided by the Federal Energy Regulatory Commission, Demand Response (DR) is defined as "Changes in electric usage by end-use customers from their normal consumption patterns in response to changes in the price of electricity over time, or to incentive payments designed to induce lower electricity use at times of high wholesale market prices or when system reliability is jeopardized".

Fig. 6.1 Energy
performance classes for
buildings according to EN
15217

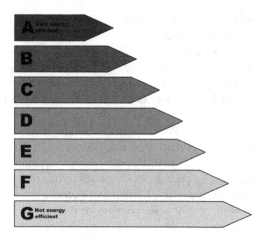

Class A: High energy performance BACS and TBM systems;
Class B: Advanced BACS and TBM systems;
Class C: Standard BACS;
Class D: Non-energy efficient BACS.

The BACS efficiency classes refer only to the installed BACS and TBM systems and not to the building as a whole, as they are not correlated to the energy classes defined by the European Standard EN 15217.

The latter energy classes are indeed referred to the energy consumption per unit of volume or surface of a building. While the EN 15232 standard creates four efficiency classes based on the presence of BACS and TBM systems.

The energy performance classes of a building are defined by the technical standard EN 15217. These classes, on the basis of the different specific energy consumption levels of the building, vary from G to A, as represented in Fig. 6.1.

For evaluating the energy performance of a building, the technical norms issued by CEN and summarized in the "Umbrella Document" CEN/TR 15615 can be adopted.

An interesting analysis of the implementation status in the European Countries of the EPBD recast can be found in (Andaloro et al. 2010).

The energy performance class of a residential building, in Italy, is calculated neglecting the energy consumption for cooling or lighting and taking into account only the primary energy consumption for heating and for sanitary hot water production. For this reason, this certification process is quite unfeasible for Southern Italy areas, where a large amount of energy is required for cooling during summer.

For calculating the amount of energy consumed, the Umbrella Document indicates the technical standard EN ISO 13790 (2008). In Italy, the standards UNI TS 11300 (2008a, b), based on the above-mentioned standard, can be used for the same purpose.

Table 6.1 Limits for EPI for class C buildings as function of the climatic zone and of the ratio S/V

EPI$_L$ (kWh/m^2)

Ratio S/V (m^2/m^3)	Climatic zone									
	A	B	C		D		E		F	
	≤600 HDD	≤601 HDD	≤900 HDD	≤901 HDD	≤1,400 HDD	≤1,401 HDD	≤2,100 HDD	≤2,101 HDD	≤3,000 HDD	>3,000 HDD
≤0.2	8.5	8.5	12.8	12.8	21.3	21.3	34	34	46.8	46.8
≥0.9	36	36	48	48	68	68	88	88	116	116

The application of these standards is not easy and is beyond the scope of this chapter, although an example of the application of the standards UNI TS 11300 to a case study can be found in (Tronchin and Fabbri 2012).

The above-recalled standards provide techniques for the calculation of the specific energy use for heating and hot water production of a building. This specific energy is the ratio of total energy used over the total built-up area/volume. This ratio is named Energy Performance Index (EPI) and is compared to a limit numerical value EPIL, devised by the Italian Legislative Decree 311/06. This comparison allows determining the *energy performance class* of a building.

The values of EPIL are given by the Legislative decree 311/06 as a function of the climatic zone in which the building is located, of the Heating Degree-Days (HDD) defined by the norm ISO 15927-6 (2007) and of the ratio S/V, where S is the total dispersion surface and V is the gross volume of the building.

These values are listed in Table 6.1.

As both the EPI and the EPIL are known, the energy performance class is found according to Table 6.2.

6.3 Active Measures for Energy Efficiency: BAC Efficiency Class in Residential Building

The European Standard EN 15232 defines the BAC efficiency classes. The standard presents two methods to evaluate the effects of building automation and management functions on the energy performance of a building:

- the "detailed method", requiring a complete knowledge of the characteristic of the building and of the installed lighting, and HVAC systems;
- the "BAC Factors method", allowing a quick but less accurate estimation of the impact of the BACS and TBM systems on the energy performance of the building in a reference period of 1 year.

Here, only the BAC Factors method is taken into consideration. According to this method, the influence of BAC and TBM systems on the energy performance of buildings is evaluated using two different numerical factors, called respectively "BAC factor for thermal energy" and "BAC factor for electrical energy". These

Table 6.2 Table for the calculation of the energy performance class of a residential building according to the DLgs 311/06

EPI (kWh/m^2)	Energy performance class
EPI < 0.25·EPIL+9	A+
0.25·EPIL+9 ≤ EPI < 0.50·EPIL+9	A
0.50·EPIL+9 ≤ EPI < 0.75·EPIL+12	B
0.75·EPIL+12 ≤ EPI < 1.00·EPIL+18	C
1.00·EPIL+18 ≤ EPI < 1.25·EPIL+21	D
1.25·EPIL+21 ≤ EPI < 1.75·EPIL+24	E
1.75·EPIL+24 ≤ EPI < 2.50·EPIL+30	F
EPI ≥ 2.50·EPIL+30	G

Table 6.3 BAC efficiency factors for thermal and electric energy for residential buildings

Single family houses, apartment block and other residential buildings	A	B	C	D
Thermal energy BAC efficiency factor $f_{BAC,hc}$	0.81	0.88	1.00	1.10
Electric energy BAC efficiency factor $f_{BAC,e}$	0.92	0.93	1.00	1.08

factors quantify differently the effects on thermal energy savings and on electrical energy savings.

They are calculated by comparing the yearly energy consumption of a given technical installation (ventilation, lighting, etc.) of a reference edifice (in BAC class C) with the consumption of the same installation evaluated in the same boundary conditions (occupation time, load profile, weather, solar irradiation, etc.) after the installation of a Building Automation and Control system for each of the four different classes (A, B, C, D). The BAC efficiency factors are differently evaluated based on the use of the buildings.

The BAC efficiency factors for thermal energy (heating and cooling systems) $f_{BAC,hc}$ and for electrical energy $f_{BAC,e}$ for residential buildings are reported in Table 6.3 and are taken from the technical standard EN 15232.

In Fig. 6.2 the calculation methodology is detailed (Ippolito et al. 2014), when using the BAC factors method to evaluate the reduction of energy consumption of a building when the BAC efficiency class is improved from a starting class 1 to a better class 2.

In Italy the technical guide CEI 205-18 (2011) provides interesting tips for the application of EN 15232.

6.3.1 Calculation of the Electric Energy Consumption

In recent years, various methodologies have been proposed to simulate the daily load power consumption in residential homes. Many of these techniques

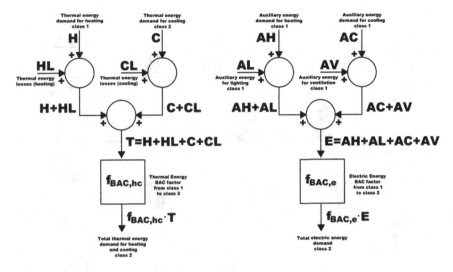

Fig. 6.2 Application of the BAC factors method

(Ala et al. 2008; Campoccia et al. 2008; Capasso et al. 1994) share a probabilistic approach allowing the construction of the daily profile starting from the knowledge of social, economic and demographic elements. Several probability functions cover the close relationship existing between the demand of residential customers and the psychological and behavioural factors that are typical of the inhabitants of the household; the models make use of such functions through a Monte Carlo extraction process. Here, the daily power profile of the test house is found according to the bottom–up approach defined in Ala et al. (2008) and implemented in the tool SirSym-Home developed by the Department of Energy Information technology and mathematical models of the University of Palermo, Italy, DEIM, within the National Research Project SIRRCE (2010).

The tool allows the calculation of the yearly energy consumption starting from the daily power profile of the building, influenced by:

- the number of inhabitants of the house;
- the period of the year (winter or summer season);
- the different working cycles of some devices (electric oven, dishwasher, washing machine, etc.).

Given the unpredictability of these factors, to proceed with the calculation of an average daily profile it is necessary to implement a Monte Carlo approach.[2]

---

[2] Monte Carlo methods are a broad class of computational algorithms that rely on repeated random sampling to obtain numerical results; typically one runs simulations many times over in order to obtain the distribution of an unknown probabilistic entity.

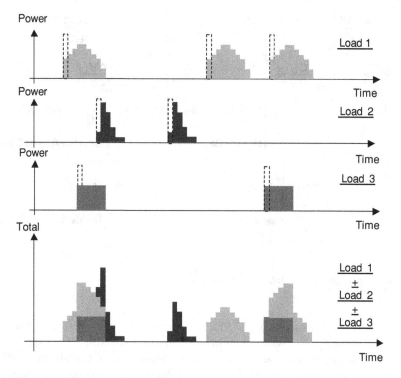

Fig. 6.3 Construction of the daily load profile

In Fig. 6.3 an example of the process of construction of the daily power profile implemented by SirSym-Home[3] is represented.

In the example proposed, the rectangles drawn with dotted lines represent the times at which the three loads are switched on during the 24 h of a typical day. These switching times are determined through a Monte Carlo extraction process considering the usage probability distributions of the three loads.

The shapes drawn in magenta, blue and green in the first three graphs represent the load profiles. The fourth graph represents the total daily load profile given by the superposition of the three single load profiles.

6.4   Economic Analysis

BACS and TBM systems are still unknown to the most part of the residential users. The limited commercial widespread of these systems is one of the most important reasons for their very high purchase and installation costs.

---

[3] SyrSim-home is a software that implements the Monte Carlo method to build the power consumption diagrams for residential uses and automated loads control.

In this situation, an economic analysis, performed by calculating the cash flow and the pay-back-period (PBP) of the investment for the installation of a BACS, becomes a very important tool for making plausible hypotheses on their commercial widespread in the medium-term.

The cash flows depend on several factors such as the initial thermal and electric demand of the house, the electricity purchase cost, the cost of the fuel for the heating system, the kind of HVAC system, the maintenance and management costs.

All these factors can be translated into cash flows Ct* by means of the following equation obtained by adding algebraically all the costs Ci and all the profits Pi related to the generic t-th year:

$$C_t^* = \sum_i P_{i,t} - \sum_i C_{i,t} \qquad (6.1)$$

In order to carry out a realistic analysis, the cash flows are annualized using the classical expression derived by Feibel (2003):

$$C_t = \frac{C_t^*}{(1+i)^t} \qquad (6.2)$$

where i is the inflation rate.

Equation (6.2) allows to obtain the equivalent present value of the cash flow of the t-th year.[4]

6.5   Tools for Extensive Evaluations of Energy Efficiency Due to Passive Measures: The "Passivhaus" Standard

In the above sections, the regulatory framework justifying the approach for the evaluation of the economic impact of some active measures for improving energy efficiency have been presented. In this section, the regulatory framework and scientific background for the economic evaluation of the impact of passive measures is presented. Tools to make such assessments can be found within the Italian standard UNI TS 11300 and within the Italian regulatory framework from the AEEG (Authority for Electrical Energy and Gas), but since they derive from extensive studies, they can easily used in other national contexts showing similar climatic features. In the last 10 year in North Europe Countries, especially in Germany, a great interest rises towards the constructive standard of the so-named Passivhaus. A Passivhaus is a building able to assure comfort conditions during winter period, without a traditional heating system. In order to attain this result, the yearly energy request for heating must not overcome 15 kWh/m^2 year.

---

[4] For example, if i = 5 %, a cash flow $C_t^* = 1,000$ € produced at the 10th year is equivalent to 614 € at year 0.

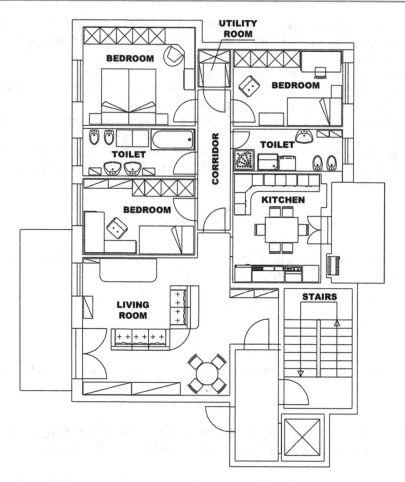

Fig. 6.4 Floor plan of the test single-family house with indication of electrical power supplied devices

The Passivhaus buildings require around 85 % less energy for heating as compared to a standard house built according to the requirements set by the Italian law n. 10/1991.[5] The total primary energy demand in these houses indeed, including electrical appliances, is limited to 120 kWh/m^2 year. As recalled in the preceding chapter, the Passive-On project was aiming at the extension of the Passivhaus standard to Euromediterranenan climates, where the need for cooling is much more important that the need for heating. The passive design tries to maximize the thermal and environmental benefits that may arise by means of the attentive analysis of the thermal performance of the building components and

---

[5] The Law n. 10/1991 is in Italy the fist law imposing energy savings and installation of renewable energy sources in buildings.

Table 6.4 EPI limits for the determination of the energy performance class of the test house

Energy performance class	Lower limit of the EPI (kWh/m^2)	Higher limit of the EPI (kWh/m^2)
A+	0	16.05
A	16.05	23.10
B	23.10	33.15
C	33.15	46.20
D	46.20	56.25
E	56.25	73.35
F	73.35	100.5
G	100.5	–

systems, so as to minimize the heat losses during winter and the thermal gains during summer. A purely passive project would exclude any mechanical means (i.e. pumps, ventilation systems,…), but this would result not appropriate in many cases, since the consideration of electrical or mechanical appliances is normally desirable to allow the passive elements to work correctly and efficiently.

6.6 Case Study

The test house is located in a Euro Mediterranean area in the city of Palermo, Sicily Italy, and is a single-family house. In what follows according to the regulatory and scientific framework described above, the economic evaluation about the impact of both active and passive measures is carried out. A further example can be found in (Riva Sanseverino et al. 2013).

6.6.1 Active Measures

The test house is reported in Fig. 6.4. It is a medium size Italian house of about 140 m^2 floor area, indoor and outdoor lighting system and a conditioning system based on fan-coil units, hot or cold water are provided by a centralized thermal/cooling system. The house is a cottage located in a Mediterranean city, Palermo, inside the climatic zone B with HDD = 751. The ratio S/V for the house is 0.3. The EPIL for the case study, calculated from Table 6.1 is 15.1 kWh/m^2 per year.

In Table 6.4, are reported the ranges of the EPI for each energy performance class for the case study, calculated on the basis of Table 6.2.

The characteristics of the supply are:

- Single-phase system;
- Rated Voltage and frequency: 230 V/50 Hz.
  The electric loads in the house are listed in Table 6.5.

Table 6.5 List of the electric loads of the test house

Category	Description	Electric power request during normal operation (W)	Stand-by consumption (W)
Lighting system	Entrance lighting	15	0
	Living room lighting	90	0
	Kitchen lighting	40	0
	Corridor lighting	30	0
	1st toilet lighting	29	0
	2nd toilet lighting	29	0
	1st sleeping room lighting	40	0
	2nd sleeping room lighting	40	0
	3rd sleeping room lighting	40	0
	Outdoor lighting	80	0
	Utility room lighting	18	0
Appliances	Polisher	1,000	0
	Hair dryer	1,200	0
	Electric shaver	40	0
	Electric iron	1,200	0
	Toaster	500	0
	Electric oven	1,600	0
	Microwave oven	1,000	3
	Fridge-freezer	260	2
	Dishwasher	2,700 (max)	0
	Washing machine	2,000 (max)	0
	Exhaust fan	170	0
	Electric dryer	1,300	0
Entertainment system	PC	100	5
	HI-FI	50	9
	TV, VCR and DVD player	300	21
	Door phone	40	25
Others	Electric storage water heating	1,200	0

Table 6.6 Electrical energy consumption of the test house for lighting and ventilation in the base case

Electrical energy consumption winter period (kWh/year)	Electrical energy consumption summer period (kWh/year)	Total electric energy consumption (kWh/year)
1,255	1,170	2,425

Table 6.7 Thermal energy consumption of the test house varying the energy class

Energy performance class	Energy for heating (kWh/year)	Energy for cooling (kWh/year)	Total thermal energy (kWh/year)
A	1,015	1,523	2,538
B	2,121	1,891	4,012
C	3,494	2,310	5,804
D	5,223	4,701	9,924
E	7,134	5,109	12,243
F	10,123	6,533	16,656
G	15,470	7,868	23,338

Table 6.8 Electric energy savings for a building upgraded from BAC efficiency class D to A

Total electric energy consumption class D (kWh/year)	Total electric energy consumption class A (kWh/year)	Energy savings (kWh/year)
2,425	2,062	363

Table 6.9 Thermal energy savings for a building upgraded from BAC efficiency class D to A

Original energy performance class	Tot thermal energy BAC class D (kWh/year)	Tot thermal energy BAC class A (kWh/year)	Energy savings (kWh/year)
A	2,538	1,878	660
B	4,012	2,969	1,043
C	5,804	4,295	1,509
D	9,924	7,344	2,580
E	12,243	9,060	3,183
F	16,656	12,325	4,331
G	23,338	17,270	6,068

Each electric load is characterized by a load profile. The load profile is the representation of the absorbed active power and of the related power factor, versus time.

In the present study, according to the common practice for residential loads, loads power factors have been considered constant in time and equal to 0.9.

Table 6.10 EPI and energy performance class before and after the application of BACS and TBM

| Starting energy performance class | Energy for hot water production (kWh/year) | BAC efficiency class D | | BAC efficiency class A | | |
		Energy for heating (kWh/year)	EPI (kWh/ m² year)	Energy for heating (kWh/year)	EPI (kWh/ m² year)	Ending energy performance class
A	1,704	1,015	19.42	751	17.54	A
B	1,704	1,821	27.32	1,570	23.38	B
C	1,704	2,913	32.98	2,157	27.57	B
D	1,704	5,223	49.48	3,865	39.78	D
E	1,704	7,134	63.13	5,279	49.88	E
F	1,704	10,123	84.48	7,491	65.68	E
G	1,704	15,470	122.67	11,448	93.94	F

Table 6.11 Components to install in the test house for turning the BAC efficiency class from D to A

Temperature control system	Lighting control system	Shutter control system	Central control system
Temperature probes in every room	Movement and lighting sensors in every room	Controls, displays, actuators for combined control of lighting, shutters, curtains and HVAC	Central unit Power supply
Temperature central unit	Dimmers		PC management software
Magnetic contacts and related interfaces for the detection of the open or closed position of doors and windows			
External temperature probe			
Actuators for fan-coil units			

The starting BAC efficiency class is D because no TBM or BACS system is installed in the house.

The BAC factors for thermal energy and for electrical energy to consider for the passage of the BAC efficiency class from D to A are found from Table 6.3:

$$f_{BAC,hc} = 0.81/1.10 = 0.74$$
$$f_{BAC,e} = 0.92/1.08 = 0.85$$

Table 6.12 Costs for the purchase and the installation of the components of the BAC system of the test house

Component	Unit price (€)	Quantity	Total price (€)
Power supply	122.00	1	122.00
Contact interface	38.70	8	309.60
Temperature probes	157.50	8	1260.00
Magnetic contacts	2.70	8	21.60
Movement and lighting sensor	27.00	16	432.00
Dimmer	42.75	5	213.75
Actuators for shutters	45.00	8	360.00
Actuators for fan-coils	63.00	8	504.00
Control/actuators for lighting	45.00	8	360.00
Display for scenario	300.00	1	300.00
Central unit	1500.00	1	1500.00
Bus	150.00	1	150.00
Installation, general costs and gain of the contractor			1512.00
Total			6500.00

Table 6.13 Economic savings obtainable for the test house passing from the BAC efficiency class D to the BAC efficiency class A for each energy performance class according to EN 15217

Starting energy performance class	Economic savings (€/year)
A	119.90
B	149.46
C	173.76
D	268.08
E	314.61
F	403.14
G	537.21

In Table 6.6, the electrical energy consumption for lighting and ventilation of the test house in the base case calculated according to (Italian Technical standard, UNI TS 11300) using the tool SirSym-Home are reported.

For simplifying the calculations, the standard year has been divided into two time periods: summer period (from the 1st of April to the 30th of September) and winter period (from the 1st of October to the 31st of March).

In Tables 6.8 and 6.9, the energy savings obtained improving the BAC efficiency class from D to A and calculated using the BAC factors method are

Table 6.14 Cash flows for every building energy class (values in €)

	A	B	C	D	E	F	G
0	−6,500	−6,500	−6,500	−6,500	−6,500	−6,500	−6,500
1	−6,380	−6,351	−6,326	−6,232	−6,185	−6,097	−5,963
2	−6,258	−6,198	−6,149	−5,959	−5,865	−5,686	−5,415
3	−6,133	−6,043	−5,968	−5,680	−5,538	−5,267	−4,857
4	−6,006	−5,884	−5,784	−5,396	−5,204	−4,839	−4,287
5	−5,877	−5,723	−5,596	−5,106	−4,864	−4,404	−3,707
6	−5,744	−5,558	−5,405	−4,811	−4,517	−3,959	−3,115
7	−5,610	−5,390	−5,210	−4,509	−4,164	−3,506	−2,511
8	−5,472	−5,219	−5,011	−4,202	−3,803	−3,045	−1,896
9	−5,332	−5,044	−4,808	−3,889	−3,436	−2,574	−1,268
10	−5,189	−4,866	−4,601	−3,570	−3,061	−2,094	−628
11	−5,044	−4,685	−4,390	−3,244	−2,679	−1,604	**24**
12	−4,895	−4,500	−4,175	−2,912	−2,290	−1,105	**689**
13	−4,744	−4,311	−3,955	−2,574	−1,893	−596	**1,367**
14	−4,590	−4,119	−3,732	−2,229	−1,488	−77	**2,059**
15	−4,432	−3,923	−3,504	−1,877	−1,075	**452**	**2,764**
16	−4,272	−3,723	−3,271	−1,518	−654	**991**	**3,483**
17	−4,108	−3,519	−3,034	−1,153	−224	**1,541**	**4,216**
18	−3,942	−3,311	−2,792	−780	**213**	**2,102**	**4,963**
19	−3,771	−3,099	−2,546	−399	**659**	**2,674**	**5,725**
20	−3,598	−2,883	−2,295	−12	**1,114**	**3,257**	**6,502**
21	−3,421	−2,662	−2,038	**384**	**1,578**	**3,852**	**7,294**
22	−3,241	−2,438	−1,777	**787**	**2,051**	**4,458**	**8,102**
23	−3,057	−2,208	−1,511	**1,198**	**2,534**	**5,076**	**8,925**
24	−2,870	−1,975	−1,239	**1,617**	**3,025**	**5,706**	**9,765**
25	−2,679	−1,737	−962	**2,044**	**3,527**	**6,348**	**10,621**
26	−2,484	−1,494	−680	**2,479**	**4,038**	**7,003**	**11,494**
27	−2,285	−1,246	−392	**2,924**	**4,559**	**7,671**	**12,384**
28	−2,083	−994	−98	**3,376**	**5,091**	**8,352**	**13,292**
29	−1,876	−736	**201**	**3,838**	**5,633**	**9,047**	**14,217**
30	−1,666	−474	**506**	**4,309**	**6,185**	**9,755**	**15,160**

Fig. 6.5 Stratigraphy of the wall including the thermal coating

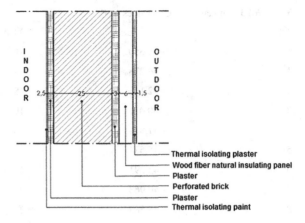

Thermal isolating plaster
Wood fiber natural insulating panel
Plaster
Perforated brick
Plaster
Thermal isolating paint

Table 6.15 Costs for the purchase and the installation of the new windows and for the realization of the thermal coating

Component	Unit price [€]	Quantity	Total price [€]
Thermal break windows	11000.00	1	11000.00
Thermal coating	20000.00	1	20000.00
Preparation of wall and roof	7250.00	1	7250.00
Painting of the external walls	3100.00	1	3100.00
Installation, general costs and gain of the contractor	15000.00	1	15000.00
Total	56350.00		56350.00

reported. In particular, such values have been obtained by multiplying the values in Tables 6.6 and 6.7 for $f_{BAC,e}$ and $f_{BAC,hc}$, respectively.

The application of BACS and TBM allows the reduction of the total thermal and electrical energy of the test house and, consequently, influences the energy performance class of the house. Considering both the energy for heating and the energy for hot water production, in Table 6.10, are reported the EPI and the energy performance class before and after the application of the class A BAC and TBM functions.

From Table 6.10 it is easy to see that the utilization of BACS and TBM systems produces an upgrade of the energy performance class of the building.

Table 6.11 shows the systems and components to be installed for improving the BAC efficiency class of the test house from D to A.

In Table 6.12 the costs for the purchase and the installation of such components and systems are reported.

The costs have been obtained as a result of a market analysis involving the most important producers of monitoring, control and automation systems in Europe.

The economic analysis is performed considering a period of 30 years.

The cash flows are annualized considering an inflation rate equal to 2.0 %.

Table 6.16 Annualized CCF (values in €)

Year	CCF	Year	CCF
0	−56,350	16	−17,843
1	−54,278	17	−15,015
2	−52,165	18	−12,133
3	−50,011	19	−9,193
4	−47,814	20	−6,196
5	−45,574	21	−3141
6	−43,291	22	−25
7	−40,963	23	**3,151**
8	−38,589	24	**6,390**
9	−36,168	25	**9,693**
10	−33,700	26	**13,060**
11	−31,184	27	**16,493**
12	−28,618	28	**19,994**
13	−26,002	29	**23,563**
14	−23,335	30	**27,202**
15	−20,615		

Moreover the electricity and natural gas costs are assumed to increase of 4.0 % each year.

Finally, in Table 6.13 the economic savings obtainable for the test house, varying the energy class from A to G are reported.

For the calculations, the electrical energy cost has been assumed equal to 0.19 €/kWh and the natural gas cost has been assumed equal to 0.091 €/kWh (EUROSTAT 2012). The economic savings can be very different from Country to Country, since they depend on climatic conditions but also on the local energy price. In particular, in Italy the latter is quite high as compared to other European Countries.

In Table 6.14 the annualized cash flows are reported as the energy class of the test house varies from A to G.

In the examined case the Payback Period (PBP) ranges from 11 years if the starting energy performance class is G to more than 30 years if the starting class is A. Nevertheless it is very important to underline that these values must not be considered as universally valid.

Indeed, the PBP as well as the cash flows depend strongly on the number, on the kind and on the market cost of the components of the BAC system installed, on the actual energy savings that can be attained per year depending on the kind of thermal/cooling system installed and on the daily management of the building.

For example, lower PBP are obtainable if a diesel boiler is installed in the test house instead of a natural-gas boiler, because the diesel cost in higher. Therefore, the analysis carried out gives only a very general indication about the convenience of the installation of BACS and TBM systems.

On the other hand, in this study, the increased market value of the house due to the improvement of its energy performance class has been neglected. A general conclusion can be drawn from the results of the calculations: the installation of a BACS system is as more convenient as higher it is the energy (electrical and thermal) consumption of the building and as lower is its initial energy class according to EN 15217.

6.6.2 Passive Measures

For the same test house, the improvement of the energy performance due to the use of some passive measures as suggested by the Passive House standard is evaluated. The measure to improve the sealing of the building using a building envelope is considered. This measure can be actuated realizing a thermal coating as the one represented in Fig. 6.5 and by replacing the old windows.

In this example, the house is supposed to have originally an Energy Performance Class G with a mean global transmittance $U = 2.775$ W/m^2 K.

After the realization of the thermal coating and the replacement of all the existing windows (originally with single glass and PVC frame) with new thermal break windows with double low emission glasses, the mean global transmittance becomes $U = 0.410$ W/m^2 K. The new Energy Performance Class is thus C. The yearly energy saving evaluated according to EEN 9/2011 is 19,043 kWh/year. Considering the same electricity and natural gas price of the previous example, the yearly economic saving obtained is 2072.23 €.

In Table 6.15 the costs for the purchase and the installation of the new windows and of the thermal coating are reported.

In Table 6.16 the annualized cumulated cash flows (CCF) due to the economic savings are reported.

In the examined case the Payback Period (PBP) is 23 years.

6.7 Conclusions

The case study presented shows that:
- the installation of a BACS or TBM system is as more convenient as higher it is the energy (electrical and thermal) consumption of the building and as lower is its initial energy class;
- the installation of a BACS or TBM system can improve the energy performance of a building and upgrade its energy class according to EN 15217.

Such general conclusions are influenced by many factors, among which the type of energy appliances and the type of heating/cooling system installed in the house.

However it is clear that the presence of BAC and TBM systems influences the energy performance class and should thus affect the energy certification process as well as the market value of buildings.

Also the application of passive measures can improve highly the energy efficiency class of the house but with very high costs and higher PBPs.

References

AEEG Deliberation EEN 09/2011, Aggiornamento, mediante sostituzione dell'Allegato A alla deliberazione dell'Autorità per l'energia elettrica e il gas 18 settembre 2003, n. 103/03 e successive modifiche ed integrazioni, in materia di Linee guida per la preparazione, esecuzione e valutazione dei progetti di cui all'articolo 5, comma 1, dei decreti ministeriali 20 luglio 2004 e s.m.i. e per la definizione dei criteri e delle modalità per il rilascio dei titoli di efficienza energetica

Ala G, Cosentino V, Di Stefano A, Fiscelli G, Genduso F, Giaconia GC, Ippolito MG, La Cascia D, Massaro F, Miceli R, Romano P, Spataro C, Viola F, Zizzo G (2008) Energy management via connected household appliances, 1st edn. McGraw-Hill, Milano

Andaloro APF, Salamone R, Ioppolo G, Andaloro L (2010) Energy certification of buildings: a comparative analysis of progress towards implementation in European countries. Energy Policy 38:5840–5866

Balijepalli M, Pradhan K (2011) Review of demand response under smart grid paradigm. In: IEEE PES innovative smart grid technologies

Campoccia A, Riva Sanseverino E, Zizzo G (2008) A Monte Carlo approach for a study on the impact of the domestic installation of small PV and thermal solar systems on the grid. In: 7th world energy system conference WESC 2008, Romania, pp 1–6

Capasso A, Grattieri W, Lamedica R, Prudenzi A (1994) A bottom–up approach to residential load modelling. IEEE Trans Power Syst 9–2:957–965

CEN/TR 15615, Explanation of the general relationship between various European standards and the energy performance of buildings directive (EPBD)—umbrella document

Directive 2002/91/EC of the European Parliament and of the Council of 16 December 2002 on the energy performance of buildings

Directive 2009/72/EC of the European Parliament and of the Council of 13 July 2009 concerning common rules for the internal market in electricity

Directive 2010/31/EU of the European Parliament and of the Council of 19 May 2019 on the energy performance of buildings

Directive 2012/27/EU of the European Parliament and of the Council of 25 October 2012 on energy efficiency, amending Directives 2009/125/EC and 2010/30/EU and repealing Directives 2004/8/EC and 2006/32/EC

European Technical Standard EN ISO 13790 (2008) Energy performance of buildings—calculation of energy use for space heating and cooling, 1st edn. CEN, Brussels

European Technical Standard EN 15217 (2007) Energy performance of buildings—methods for expressing energy performance and for the energy certification of buildings, 1st edn. CEN, Brussels

European Technical Standard EN 15232 (2012) Energy performance of buildings—impact of building automation, control, and building management, 2nd edn. CEN, Brussels

European Technical Standard EN ISO 15927-6 (2007) Hygrothermal performance of buildings—calculation and presentation of climatic data—Part 6: Accumulated temperature differences (degree-days), 1st edn. CEN, Brussels

EUROSTAT (2012) Electricity and natural gas price statistics on May 2012

Feibel BJ (2003) Investment performance measurement, 1st edn. Wiley, New York

Ippolito MG, Riva Sanseverino E, Zizzo G (2014) Impact of building automation control systems and technical building management systems on the energy performance class of residential buildings: an Italian case study, energy and buildings n. 69, pp 33–40

Italian Law n.10/1991 (1991) Norme per l'attuazione del Piano energetico nazionale in materia di uso razionale dell'energia, di risparmio energetico e di sviluppo delle fonti rinnovabili di energia

Italian Legislative Decree n.311/06 (2006) Disposizioni correttive ed integrative al decreto legislativo 19 agosto 2005, n. 192, recante attuazione della direttiva 2002/91/CE, relativa al rendimento energetico nell'edilizia

Italian Technical Standard CEI 205-18 (2011) Guide to building automation identification of functional block diagrams and estimation of related energy savings, 1st edn. CEI, Milano

Italian Technical Standard UNI TS 11300-1 (2008) Energy performance of buildings—part 1 calculation of energy use for space heating and cooling, 1st edn. UNI, Milano

Italian Technical Standard UNI TS 11300-2 (2008) Energy performance of buildings—part 2 calculation of energy primary and energy performance for heating plant and domestic hot water production, 1st edn. UNI, Milano

Passive-On Project (2013). http://www.passive-on.org

Riva Sanseverino E, Zizzo G, La Cascia D (2013) Economic impact of BACS and TBM systems on residential buildings. ICCEP 2013, Italy, pp 651–655

SIRRCE (2010) System for the residential energy optimization with summer air conditioning integration, research project, financed by the Italian Minister for the economic development with decree of the 16th of February 2010

Tronchin L, Fabbri K (2012) Energy performance certificate of building and confidence interval in assessment: an Italian case study. Energy Policy 4:176–184

Funding Energy Efficiency Measures

7

Silvia Dell'Atti

Abstract

In this chapter some difficulties for the financing of the energy efficiency measures are illustrated together with some juridical tools, which may support and facilitate the funding of the energy efficiency interventions.

7.1 Tools to Foster Energy Efficiency in Italy

As already mentioned above in this manuscript, the European Union plays a crucial role in increasing energy efficiency policies amongst its Member States and it sets the legal framework for the Italian legislation on energy efficiency. In particular, it is worth mentioning that article 194 of the Treaty on the Functioning of the European Union states that "In the context of the establishment and functioning of the internal market and with regard for the need to preserve and improve the environment, Union policy on energy shall aim, in a spirit of solidarity between Member States, to: (a) ensure the functioning of the energy market; (b) ensure security of energy supply in the Union; (c) promote energy efficiency and energy saving and the development of new and renewable forms of energy; and (d) promote the interconnection of energy networks".

Article 1 of directive No. 2012/27/EU sets out the framework for the European Union to ensure the achievement of 20 % energy efficiency target by 2020. In this context, an important pillar for achieving 20 % savings of energy consumption

S. Dell'Atti (✉)
Law Firm Macchi di Cellere Gangemi, Via G. Cuboni 12, 00197 Rome, Italy
e-mail: s.dellatti@macchi-gangemi.com

E. Riva Sanseverino et al. (eds.), *Smart Rules for Smart Cities*, Sxi 12,
DOI: 10.1007/978-3-319-06422-2_7,
© Springer International Publishing Switzerland 2014

concerns the amount of energy saved in buildings. Indeed, whereas (16) of directive No. 2012/27/EU states that "buildings represent 40 % of the EU final energy consumption [...] and Member States should establish a long-term strategy beyond 2010 for mobilizing investment in the renovation of residential and commercial buildings".

Amongst the European various legal instruments enacted by the EU to foster the energy efficiency and reach an overall objective of an energy efficiency target of saving 20 % of energy consumption we need to indicate directives No. 2006/32/EC, No. 2009/125/EC, No. 2009/28/EC, No. 2010/30/EU and No. 2012/27/EU (just to mention some of them). It is necessary to remember that generally speaking directives must be implemented in the Member States, as they are not immediately effective there. Italy has applied the above-mentioned directives by mostly adopting legislative decrees (at the time of drafting this manuscript the directive No. 2012/27/EU has not been implemented yet but just before publishing this contribution a bill of law has been submitted to the Parliament in this respect). For instance, a definition of "energy efficiency" from a legal viewpoint is included in directive No. 2006/32/EC ("a ratio between an output of performance, service, goods or energy, and an input of energy") and article 2 of legislative decree No. 115/2008 (which implements directive No. 2006/32/EC) defines "energy efficiency" as "the ratio between the results in terms of performance, services, goods and/or energy to be meant as the services rendered, and the input of energy".

A starting point of our analysis of the Italian legislation can be considered the **SEN** (*Strategia Energetica Nazionale*, literally "National Energy Strategy" as enacted by the decree dated 8 March 2013) considering that the SEN should show the path for the future laws in the energy sector (even if it is not truly a starting point, since many previous laws have tried to increase energy efficiency). First of all the energy efficiency is the **first priority** set forth in the SEN. The SEN establishes that in 2050 the private energy consumption of energy (if compared with the same in 2010) shall be reduced by a range between 17 and 26 %.

According to the SEN, the major obstacles preventing the adoption of technologies for the improvement of the energy efficiency are:

- in a non-industrial and non-public context high investments and lack of knowledge;
- in the public sector the fact that it cannot benefit from tax credit and tax deduction and the difficulty to finance such kind of intervention which would require a large use of the ESCO[1] model;

---

[1] Directive No. 2006/32/CE has introduced the regulation of ESCO in the EU. The ESCO is defined as "a natural or legal person that delivers energy services and/or other energy efficiency improvement measures in a user's facility or premises, and accepts some degree of financial risk in so doing. The payment for the services delivered is based (either wholly or in part) on the achievement of energy efficiency improvements and on the meeting of the other agreed performance criteria". This directive has been implemented in Italy by legislative decree No. 115/2008 which has introduced the concept of the ESCO (the Energy Service Company) (which incurs in a certain degree of financial risk) and ESPCo (the Energy Service Provider Company) (which does not incur any risk).

- in the industrial sector the lack of internal knowledge and the lack of appeal for investments, which have a long-term payback.

In order to improve the level of energy efficiency the SEN focuses on:

- a longer applicability of the tax deduction/credit system;
- other forms of public support (by way of example in the public sector the so called "*conto termico*" and the usage of the EU funds);
- a set of rules, which should facilitate better energy efficiency (inter alia the rules set forth by directive No. 2002/91/CE and directive No. 2010/31/EU for new buildings, the energy certification/audit, inspection on the plants, the criteria for qualification of energy efficiency experts etc.);
- the system of the *certificati bianchi* ("white certificates") for interventions in the industrial sector.

Furthermore, according to the SEN, actions for improvements made on energy efficient buildings open the door to a reassessment of the planning and management of the cities taking into consideration that 70 % of the energy is consumed in the cities.

The law ascribes a leading role in the achievement of energy efficiency to public entities (article 12 and following of legislative decree No. 115/2008[2]).

The public bodies:

(1) must identify a "responsibility for achieving the goals of energy efficiency";
(2) in relation to the public buildings,

- shall use financial instruments for energy savings (in other words they must use any financial means made available to both public and private bodies in order to cover wholly or partially the costs of the projects) in case of realization of requalification interventions;
- must carry on energy audits in case of a heating systems restructuring or in case of a restructuring of the building and they must obtain the energy certifications for buildings according to the thresholds of the floors;
- in case of new buildings or restructuring must apply the rules on energy performance set forth by legislative decree No. 192/2005 (which implements Directive 2002/91/EC on the energy performance of buildings) and as further amended by the law decree No. 63/2013 (implementing Directive No. 31/2010/EU) and must pursue the objective of "quasi zero-energy buildings" (please see article 4-bis of legislative decree No. 192/2005), promoting the recourse to ESCO schemes, the recourse to public-private partnership and financing with third parties (the so called "third-party financing contracts"), exploiting the funds of the *Cassa conguaglio per il settore elettrico* (please see art. 4-ter of legislative decree No. 192/2005 and article 22 of legislative decree No. 28/2011). Again, in order to facilitate the funding of these

---

[2] The legislative decree No. 115/2008 implements directive No. 2006/32/EC on energy end-use efficiency and energy services and repealing Council Directive No. 93/76/EEC. Please also see whereas (15) of directive No. 2012/27/EU according to which the public sector constitutes an important driver to stimulate market transformation towards more efficient products, buildings and services.

initiatives, the public bodies shall use the model of energy performance contracts for the increase of the energy performance which must guarantee the results of energy savings and the latter must be identified and measured;

- must opt for equipment and devices having reduced energy consumption, when the public bodies buy devices.

As indicated in the SEN the Italian legislator has taken three different ways in order to achieve the goal of energy efficiency increase:

- the tax deduction;
- the *conto termico* incentive;
- the white certificates.

7.1.1 The Difficulties for Investments in Energy Efficiency and the Pending Questions

Finding the resources not to consume more but incentivize and finance the reduction of the consumption seems to be inconsistent with the way our society works.

The sponsor of an energy efficiency initiative may encounter difficulties in finding the resources to finance the energy efficiency initiative and to persuade a bank to finance its initiative. Firstly, in point of fact, the use of funds for an energy efficiency initiative may divert resources from the client's core business. Secondly, converting savings into cash flow in order to enhance the bankability of energy efficient initiatives is not a simple procedure.

The banks, on the other hand, may question (1) if the prospective client is reliable, (2) how long the client will carry on its business in a way which is similar to the current one, (3) to what extent this client will benefit from the energy efficiency initiative, (4) if during the amortization period of the energy efficiency initiative the client will have enough cash flow to pay the energy efficiency services and (5) if the client has a track record in the energy sector. Furthermore, if the prospective client is not an "energy professional", the bank may also question the ability of the client to manage an investment in a profitable way, as it may not conform to the client's core business. In broad terms, the bank may question the bankability of the prospective client.

Moreover, which kind of security interests may be offered? Obviously there could be limitation on the creation of security interests over client's movable and immovable assets since it is likely that the client's assets are already mortgaged and/or already utilized as a guarantee for other loans granted to the same client.

If the energy efficiency investments benefit from incentives, unfortunately it is not possible to create a security interest over the mechanism of the *conto termico* and over the white certificates (so called TEE) (please see below).

It is not less complicated for the public entities, which may encounter difficulties to invest, to manage complex tenders, to structure complex contracts also considering that, after structuring the tender, after awarding the tender and after structuring the contract, the funds to the energy efficiency project may still be missing.

Finally, there are queries which may also remain unanswered such as whether the energy efficiency may create a partnership between public and private bodies in order to optimize investments which may encourage long term investments towards adequate rate of return and how during this time of economic and financial crisis the public sector can play a leading role which is ascribed to it by the legislation in force.

7.1.2 In Brief the Bankability of Tax Deductions and Other Recent Strategies

Regarding the first instrument which is stated by Italian legislation (lately provided for by articles 14, 15 and 16 of law decree No. 63/2013,[3] which extend with some amendments the benefit of previous law provisions dated back to 2006[4]), the tax deductions connected to energy efficient interventions (such as the installation of photovoltaic modules, of more efficient heating plants, interventions to put a cap on building etc.) vary from 55 to 65 % limited to a maximum amount. A certain number of requirements must be met in order to benefit from the tax deductions:

- the energy efficient intervention must be paid within 31 December 2014 (or 30 June 2016 if the interventions are made on the common real estate of condominiums) by bank transfer which expressly indicates the reason of the payment including the tax deduction reference;
- tax codes of the payer and payee, also an asseveration from a technician (if not included in the declaration of director of works for the end of works) or, depending on the kind of the intervention, a warranty of the producer may be required;
- within 90 days from the end of works a communication to ENEA, depending on the kind of the intervention, may also be necessary.

It is worth noting that the tax deductions connected to an energy efficiency initiative produce a tax saving and not "a cash flow". A tax deduction, as matter of fact, is not by itself destined to energy efficient initiatives financed by banks, save that these initiatives are included in more general financing plans in the context of which also the energy efficiency tax deduction is taken into consideration. By way of example, the tax deduction may be taken into account in the context of a building restructuring financing, where the banks make available loans the purpose of which is the financing of the restructuring/renovation of the building, including, amongst other items, the energy efficient intervention, but not specifically destined to finance the energy efficient initiatives solely. In the above-mentioned case, the loan documentation may include covenants on the borrower to comply with the

---

[3] Converted into law by law No. 90/2013.

[4] Please see article 1, paragraph 139 of law 27 December 2013, No. 147 named *Legge di stabilità 2014* which provides for an extension of the duration of the tax deductions.

relevant regulation on the tax deduction applicability (lately see the rules of circular letter of the Revenue Office No. 29/E dated 18 September 2013) and the financial model calculation may assume the benefit of the tax deduction. In other words, the tax deduction will be one aspect evaluated by the banks in order to ascertain the bankability of the building restructuring/renovation financing and may enhance the bankability of the project (i.e. the financing of the restructuring/ renovation of the building). In addition to the fact that the tax deduction does not generate by itself a cash flow, a further limit to the ability of the tax deduction to enhance the bankability of an energy efficient project is the lack of stability of the tax deductions or the fact that the benefit of the tax deduction is limited in time because the projects may benefit from the tax deduction only if the expenses are paid by 31 December 2014 (or by 30 June 2016 if the energy efficient initiative relates to the part of the building which form part of the condominium). The limitation in time cuts down the possibility to consider the tax deductibility in the screening of the financing of energy efficiency project because usually programming and obtaining a loan in this period of credit crunch requires a longer time and it shows the lack of planning and perspective of the Italian government and legislator (or in the worst case the short-sightedness). The above mentioned circular letter of the Revenue Office clarifies the meaning of expenses incurred by 31 December 2013 (or by 30 June 2014, in relation to the laws previously in force) based on the accounting criteria adopted by the person who is willing to benefit from the tax benefit if on an accrual basis or on a cash basis.

Finally, article 6 of law decree No. 102/2013 has provided that the *Cassa Depositi e Prestiti S.p.A.* (a company controlled by the Ministry of the Economy and Finance) may finance banks operating in Italy and granting mortgage loans destined to the purchase of the houses, the restructuring and energy efficiency actions in order to facilitate the grant of loans for the residential real estate market and in order to mitigate the effects of the economic crisis and credit crunch. At the time of writing this manuscript it is not possible to assess the impact of this new instrument and how this will expedite the refurbishment of the dwellings including the enhancement of their energy performance.

7.1.3   The Incentives for the Energy Efficiency: The Conto Termico

At the date of drafting this section the "state of the art" in Italy of the incentives for the energy efficiency are the *conto termico* and the *titoli per l'efficienza energetica* (or TEE) (called also the white certificates (*certificati bianchi*)), each of them are alternative to the other. Articles 27 and 28 of the legislative decree No. 28/2011 establish the main rules relating to the TEE and the *conto termico*. Preliminary it should be noted that the purpose of the legislative decree No. 28/2011 (implementing Directive No. 2009/28/CE) is promoting the achievement of the national goals set out under article 3 of the legislative decree No. 28/2011 to be met in 2020 (which include 17 % fuelled by renewable sources on the energy final consumption) and therefore it has not regulated in details the operation of the *conto termico*. The detailed rules have been adopted one year after the enactment of the legislative decree by Ministerial Decree dated 28 December 2012. The *conto termico* main scope is supporting the public entities to increase the energy efficiency because the public entities cannot take advantage from the tax deductions (from which the private persons benefit) while both private and public entities may avail themselves of the *conto termico* for producing energy from renewable resources or for high efficiency plants (for more details see below).

The main requirements for benefitting from the *conto termico* are:

- The interventions must be realized after 31 December 2011 and may benefit from the *conto termico* to the extent which exceeds the quota necessary to comply with the mandatory rules on the inclusion of renewable resources in the buildings of new construction and in the buildings which are object of property renovation and which are necessary for the obtainment of the issuance of the building permits (article 4, third paragraph, Ministerial Decree 28 December 2012);
- The Ministerial Decree distinguishes between the "admitted persons" (who are the beneficiary of the interventions) and the "responsible persons" (who have paid for the expense of the intervention, are entitled to receive the incentive and execute the agreement with the GSE);
- The admitted persons may be public entities and private persons (including natural persons, business enterprises, condominiums). The admitted persons may also avail themselves of the ESCO[5];
- The responsible persons are the persons who have incurred the expense for the intervention and who executes the agreement with the GSE. If the public entities or the private persons have executed a contract for the energy services, then the responsible person is the ESCO, on the grounds that it is the subject that has incurred the expense, whilst in case of financing by third parties (which occurs when the financing is made available by a bank), the responsible person is the public entity or the private person.[6]

The funds which are available for financing the *conto termico* are limited: for the private sector are capped up to €700 million, whilst the public entities may

---

[5] According to the Ministerial Decree and the Rules for the applicability of the Ministerial Decree dated 28 December 2012 released by GSE on 4 December 2013 (formerly release dated 9 April 2013) the ESCOs relevant for the *conto termico* are (i) the entities that are certified UNI CEI 11352 (ii) entities whose articles of association provide the supply of energy services and that are included in the list of Energy Service Companies (*Società di Servizi Energetici*) currently managed by GSE and (iii) the entities indicated in article 2, first paragraph of legislative decree No. 115/2008. The above mentioned entities provide energy services or any other instruments for the improvement of energy efficiency in the plants or areas of the client accepting a financial risk. If the responsible person is the ESCO, the energy contract executed by the ESCO together with any financing agreement made available by third party, which evidences the incurred expenses, must be filed with GSE.

[6] In case of leasing, the invoices must be issued to the leasing company and a copy of the leasing agreement must be delivered to the GSE S.p.A.

benefit from €200 million expenditure engagement and from the right to book in advance the incentive (up to €100 million). The public sector may obtain the *conto termico* in relation to energy efficiency interventions (such as thermic insulation of the opaque surface, replacement of transparent fixtures, replacement of existing heating plants with new and more efficient heating plants and shading systems) on buildings which are already existing, duly registered in the cadastral register and which have a plant for the regulation of the internal climate. Both the public and private sector may obtain the *conto termico* in relation to interventions on already existing buildings for plants producing thermal energy from renewable sources and high efficiency systems (such as the replacement of heating plants with new one having heat pumps, replacement of heat plants for greenhouses with a heat plant fuelled by biomass etc.). For more complex energy efficient initiatives it may be necessary to enroll in a registry in order to benefit from the *conto termico* incentive.

The *conto termico* is fixed for the whole period of its duration and shall ensure a fair remuneration of the investment and operation costs. The right to the incentive starts from the date of conclusion of the intervention, as stated in the request to be filed with *Gestore dei Servizi Energetici S.p.A.* (the "**GSE**") up to a maximum of 5 years.

The request for the incentive must be filed within 60 days from the completion of the energy efficient intervention and it must include bank transfer payment evidencing that the payment is connected to an energy efficient intervention benefitting from the *conto termico*. If the energy efficient intervention is financed by means of leasing or other kind of financing, the payment order must include details indicated in the Rules for the applicability of the Ministerial Decree dated 28 December 2012 released by GSE.

The incentive is granted after the execution of an agreement by and between the person responsible for the intervention and GSE according to contract model approved by the *Autorità per l'energia elettrica il gas e il sistema idrico* (the Italian regulatory body for the electric energy, gas and hydric system).

The *conto termico* incentive should aim at reducing the difficulties in energy efficiency initiatives by reducing the cost of the investment. Notwithstanding this goal, the form of the agreement for receiving the incentive under article 5 (schedule 10 "*facsimile scheda Contratto*" to the rules implementing the Ministerial Decree 28 December 2012 release 4 December 2013[7]) provides some limitations to the fully bankability of the agreement for receiving the incentive. More specifically, it provides that "the Responsible Person is allowed to assign the

---

[7] Please consider that the current version of the agreement for receiving the *conto termico* incentive has enhanced the possibility to finance the energy efficiency interventions. The current version of the agreement for receiving the *conto termico* incentive is a true improvement if compared to the previous release of the agreement, which provided that "the Responsible Person is not allowed to create a pledge over the incentives (matured or to be matured) or to assign the account receivables. It is also not allowed to grant a special mandate with third party or to use any kind of delegation".

receivables relating to the money indicated in the following article 6. The GSE will pay to the assignee provided that:

(a) the assignment relates to all the receivables of the assignor towards the GSE pursuant to this contract;
(b) the receivables will be assigned to a single assignee only;
(c) the deed of assignment
  - is to be signed on the same date or after the signing of this contract;
  - is to be drafted in fully compliance with the format published by GSE on its website by filling the relevant gaps;
  - shall be executed before a notary public or with signature certified by a notary public.

It is not allowed to create a pledge over the receivables (matured or to be matured under this contract or to make any assignment following to the first one from the assignee to third party).

Therefore, because of the limitations provided for by the contract to be entered by the beneficiary and the GSE, it is possible to grant as guarantee on the cash flow generated by the *conto termico* only the pledge over the bank account over which the incentive will be paid into and the assignment of the receivables arising under the *conto termico* (subject to the conditions indicated by the GSE).

7.1.4 The Incentives for the Energy Efficiency: The Titoli di Efficienza Energetica

The white certificates or *"Titoli di Efficienza Energetica"*[8]("**TEE**")(Energy Efficiency Securities) are instruments tradable which certify that energy savings in the final use of energy have been achieved by means of interventions for the energy efficiency increase.

The white certificates regime has been introduced by article 9 of legislative decree No. 79/1999, DM 20 July 2004, by legislative decree No. 115/2008, by legislative decree No. 28/2011, by Ministerial Decree 28 December 2012.

Pursuant to above mentioned laws and regulations the electric and gas distribution companies must achieve yearly targets of primary energy saving (the saving is measured as a saving of oil equivalent ton or *Tonnellate Equivalenti di Petrolio* (TEP)).[9] A TEE is equal to a saving of one TEP.

The goal of primary energy saving is made directly (by making interventions of energy savings or through the help of ESCOs) or indirectly by purchasing the TEE from third parties. The ESCOs, the major industrial and services operators which are obliged to appoint the energy manager, the gas and electric distribution

---

[8] Specific rules are applicable to high yield cogeneration plants.

[9] Originally under article 4 of the Ministerial Decree 20 July 2004, electric and gas distribution companies, which had more than 100,000 clients, were eligible and qualify for the obligation to gain the energy savings from the TEEs system; thereafter the threshold has been reduced to 50,000 clients.

companies which do not qualify for the obligation of TEEs and the companies controlled by the electric and gas distributions companies obliged to the TEEs mechanism may freely decide to be part of the TEEs mechanism. There are restrictions to the fact that the plant benefiting from TEE may benefit from other kind of incentives pursuant to article 10 of Ministerial Decree 28 December 2012.

If there are some players, which are obliged to evidence the energy savings initiatives by obtaining a number of TEEs each year, the legislator has envisaged a system where there is also purchaser of last instance (the GSE). Starting from 2017, where the quantitative goals following 2016 or other instruments for the safeguarding of the investments have not been defined, the GSE shall off take all the TEEs produced by already realized investments paying to the producer a fixed amount (equal to the average price of the transactions for the years 2013–2016 discounted of 5 % pursuant to article 4 of Ministerial Decree 28 December 2012).

The Ministerial Decree 28 December 2012 sets the saving targets for year 2013, 2014, 2015. The interventions which entitle the issuance of TEEs may be carried out by the same persons which are obliged to pursue the energy saving targets (i.e. the electric and gas energy distribution companies and their controlled companies) as well as by other persons (such as energy companies, companies which deplete energy over defined thresholds, other kind of companies or public entities which have appointed a responsible for the saving and reasonable use of energy or which are ISO 50001 certified). The persons involved in the issuance and purchase of TEEs are enrolled in the electronic Register of the TEEs, held by the *Gestore dei Mercati Elettrici S.p.A.* (the company which is the regulator of the energy markets, also the "**GME**"), whilst the GSE will apply sanctions on the electric and gas distribution companies which do not reach the targets.

According to Ministerial Decree 5 September 2011 the right to benefit from TEEs may be granted to:

- New plants (for a 10 year period or up to 15 years under certain conditions);
- Revamping of existing plants (for 5 year period).

The legal framework for the TEEs is quite similar to the legal frame of the green certificates, which may be considered a (partial) success, but specific difficulties narrow the bankability of TEE projects.

In order to facilitate the spread of financing the TEE projects, the GME has envisaged a procedure for blocking/unblocking the TEEs to allow the owners of the TEE account with the TEEs register to use, in whole or in part, the TEEs (which are available to the owner) to secure the credits of the banks financing energy efficiency projects.

The procedure is quite simple.

The owner of the TEE account shall deliver to the GME (by return receipt letter anticipated by mail) a request signed by the owner and the secured creditor for blocking any transaction of TEEs (on the market and for bilateral arrangement) according to the form approved by GME together with a statement on the authority to sign the request. The GME shall verify the completeness and formal regularity of the documentation received and within the forth business day after the receipt of the blocking request the GME shall block the TEE account. The GME provides the secured creditor with a user-id and password which however it will allow the bank to check and monitor the status of the account only. From the viewpoint of the lender this procedure entails some limitations both because the secured creditor shall get involved in the management of the guarantee and because the TEEs, which will be blocked, must be identified in advance and this implies that they must already be registered in the account. At the end each year the lender and the debtor should sign a new agreement (or an agreement confirming and containing new details) for blocking the new TEEs registered in the account and therefore in case of insolvency of the debtor it might be subject to claw back (in other words there is the risk that guarantee may not consolidate completely). Additionally, the model agreement approved by GME must be wholly reflected in the actual agreement signed by the secured lender and the debtor and therefore it cannot be customized.

7.2 The Private Talent: The Agreements with the ESCO

As already anticipated in many points of this manuscript a crucial role should be played by the ESCO (at least in the imagination of the legislator, but the reality is not gone so far yet but in the future it could spread out). We have already discussed above in this manuscript about the features of the ESCO. Contracting an ESCO is advantageous in that the ESCO may commit to finding the necessary financial resources and it commits to carrying out the energy audit, the engineering and feasibility plans, as well as it realizes, operates and manages the intervention. Generally the ESCO is the owner of the plant and of the equipment necessary to carry out the energy service. The scope of the energy contract with the ESCO is not the construction of the plant or the supply of the equipment but the supply of energy services (heating energy, equipment and plant which guarantee the achievement of the energy saving goals). The considerations of the ESCO may be the incentives, any available public benefit and the price for the energy service contract. The ESCO manages the intervention and the supply of energy, while the result of an energy consumption saving may be shared between the ESCO and the client by means of (i) the sharing of the saving as compared to the pre-intervention energy expense or (ii) a guarantee of the saving, i.e. the ESCO warrants a saving with respect to the pre-intervention expense and therefore the client pays to the ESCO an energy instalment which is equal to the pre-intervention energy intervention less a percentage of saving. In this latter case the profits for the ESCO are floating.

The client of the ESCO states the provisions of energy services in the contract with the ESCO for a medium-long term which allows the investor to recover the amortization of the investments made for delivering the energy services. The parties involved the energy service contract may freely agree on the duration of the contract, when the ownership equipment and plant will be transferred to the client, and how to share the energy saving between the client and the ESCO for the whole duration of the contract. If the energy service contract is awarded by a public entity to an ESCO according to the Authority for the Supervision of the Public Contracts of Works, Services and Supplies (resolution AVPC No. 37/2012) this falls into the category of "composite contract".

The Italian legislator has defined both the energy service contract[10] (which is the contract which establishes the supply of goods and services necessary to maintain the comfort in the buildings in compliance with the laws on the rational use of energy, on safety and environmental preservation, and which improves the procedure of transformation and use of energy) and the EPC (which is the acronym for energy performance contract[11] as the agreement between the beneficiary and the supplier of an energy efficient improvement measure where investments in that measure are paid for in relation to the agreed level of energy efficiency improvement) and which would let the ESCO be paid out by both the savings and incentives generated by the energy efficient interventions and renewable energy projects. The scope of the energy performance contract is the engineering, realization and operation of an energy efficient intervention which reduces the energy consumption. In conclusion the ESCOs offer the diagnosis, the project, the intervention for the increase of energy efficiency and the management of the intervention at no cost to the client (both a public entity client and a private client). Under the EPC the technical risk is on the ESCO and if the ESCO offers to finance the energy efficiency intervention under the third-party financing scheme (please see below), it is also exposed to the financial risk of the energy efficient interventions.

According to the third-party financing scheme in order to promote the investments in the energy efficiency a third party (which is different from the energy supplier and from the beneficiary of the energy improvement measure, but can be an ESCO or a bank) may provide the capital for the energy investment and may be repaid by charging a fee equivalent to a part of the energy savings achieved as a result of the energy efficiency improvement measure.[12] In the event that the ESCO finances the energy efficiency investments, it needs to obtain enough financial resources from its shareholders (for instance by means of agreements under which the shareholders undertake to contribute additional equity if necessary) and from banks which will assess the bankability of the ESCO on the basis of the project (i.e. the energy efficient intervention for which the ESCO needs additional resources) and on the basis of the guarantees which the ESCO and its shareholders may offer. Amongst the guarantees which the ESCO and its shareholders may create in favor of the bank lending to the ESCO, it is worth mentioning the pledge over the shares of the ESCO, the pledge over the bank accounts of the ESCO (where any incentive of the *conto termico* may be paid into), the guarantee over

---

[10] Article 16, paragraph 4 of legislative decree No. 115/2008.

[11] Article 2, letter (l) of legislative decree No. 115/2008 defines the *contratto di rendimento energetico*.

[12] Please see article 3, letter (k) of directive No. 2006/32/EC.

the white certificates[13] (if any), the special privilege pursuant to article 46 of legislative decree No. 385/1993 over the movable assets of the ESCO (which shall include the plants and equipment necessary for the energy efficiency improvement), any pledge over the accounts receivables arising from the EPC or energy service contract and any relevant contract or the assignment by way of security of the same. Furthermore, it cannot be excluded that in the next future the ESCOs may avail themselves of the possibility to issue the so called "minibonds" exploiting the new rules and tax advantages on the bonds enacted by article 32 of the law decree No. 83/2012 and by article 12 of law decree No. 145/2013 (the so called "*Destinazione Italia*" decree).

The pros of the use of the contracts with the ESCOs are inter alia that:

- it reduces the financial investment burdens for the client;
- the complete management of the investment risk is allocated on a professional who can act on the basis of his competence, expertise and his consolidated and standardized criteria;
- it can benefit from *conto termico*[14] and TEEs.

The contra are inter alia that:

- the medium-long term contract is subject to the risk of the business of the client who does not pay for the investment and may be encouraged to be opportunistic;
- if the investment involves assets which are not movable, any existing mortgage of the client may extend to the investment made by the ESCO.

A possible risk mitigation for the ESCO is the risk-sharing with the client (by means of creating a special purpose vehicle for the single project controlled by the ESCO and where the client has minority shareholder rights and an undertaking to inject additional equity under specific circumstances) and/or with the supplier of the equipment and plant and/or with the supplier of the energy and/or with a combination of sharing the risks with the client and the supplier (if the latter accepts to be paid at a later stage by executing a vendor loan or other modalities of deferred payment).

---

[13] Please see above for more details.

[14] In case of *conto termico* incentive the ESCO must enter into a contract for the energy performance with the public entities and for the interventions provided for by article 4, second paragraph of Ministerial Decree 28 December 2012. The EPC contract must comply with the requirements set out by the law provisions, lacking which the energy efficiency intervention can lose the *conto termico* incentive.

Smart Planning and Intelligent Cities: A New Cambrian Explosion

8

Maurizio Carta

Abstract

We live in the society of knowledge, creativity and innovation: true anti-cyclical factors with respect to the crisis that has overrun the traditional development protocols and that requires powerful processes of creation and spread of knowledge. The true innovation has no boundaries, it has to affect each aspect of institutions and enterprises and operates as a mutagen of society, requiring a paradigm shift. Startups, fablabs, co-workers, makers and smart citizens have given rise to a global urban movement and most cities now have a sizeable colony: a true smart ecosystem for improving social innovation. Between them they are home to hundreds of accelerators and thousands of smart places and co-working spaces, and this ecosystem must be highly interconnected and integrated in a renewed urban metabolism driven by more adequate planning paradigms and tools. The combination of technological innovation and urban planning, however, is not only instrumental and determines changes within the community and its territory too. The "Third Industrial Revolution" and the gradual implementation of e-society have made it possible to delegate an increasing number of physical and intellectual tasks, even very sophisticated, to technology. In fact, the goods and ideas produced are increasingly less tied to a scheduled place and time, in terms of quality and quantity; the workplace is no longer an independent variable and time is no longer rigidly synchronized, especially as far as the intellectual work is concerned. The spreading of sensors, smart devices, electronic networks and urban life apps has created a proper urban cyber-physical space, consisting of the constant interaction between physical components and digital networks,

M. Carta (✉)
Department of Architecture (DARCH), University of Palermo, Palermo, Italy
e-mail: maurizio.carta@unipa.it

E. Riva Sanseverino et al. (eds.), *Smart Rules for Smart Cities*, Sxi 12,
DOI: 10.1007/978-3-319-06422-2_8,
© Springer International Publishing Switzerland 2014

tangible actions and intangible feedback. Smart cities are components of a new urban organism able to rethink the development and to encourage a "creative explosion", leading smartness-based initiatives as part of a European post-metropolitan vision.

8.1 Open Urbanism Scenario

We are increasingly immersed in the society of knowledge, creativity and innovation, today universally regarded as the key to competitiveness, true anti-cyclical factors with respect to the crisis that has overrun the capitalist development protocols and which requires processes of knowledge creation, spread and replacement. It requires a constant, powerful and pervasive flow of knowledge, exchange of information, and instant evaluation about the effects of government actions. Innovation has no boundaries, it affects each and every an aspect of institutions and enterprises and operates as a "mutagen" of society, requiring a paradigm shift to whom bears the responsibility of governing under the aegis of a renewed leadership. In early 2014 *The Economist* has published a report about the rise of startups and smart communities, recognizing a new *Cambrian Explosion*: "digital startups are bubbling up in an astonishing variety of services and products, penetrating every nook and cranny of the economy. They are reshaping entire industries and even changing the very notion of the firm".[1] Startups, fablabs, makers and smart citizens have given rise to a global urban movement and most cities now have a sizeable colony: a true smart ecosystem. Between them they are home to hundreds of accelerators and thousands of smart places and co-working spaces, and all these ecosystems must be highly interconnected and integrated in a renewed urban metabolism driven by more adequate planning paradigms and tools (Carta 2014).

Today the new path ahead of the world socio-economies is to draw on the long network flows, transforming them through spatial patterns into energy for local systems. These flows, once diversified into veins of identity, generate value in the local realm to be re-entered in the large global corridors that will thus be revitalised, nurtured, characterized and differentiated. Among the challenges resulting from the connections between global and local, knowledge and skills are the most ambitious and complex. It is no longer just a matter of *Lisbon Strategy* (2000), which advises such a paradigm shift, but a shared need for hope. A hope for the

---

[1] See: A Cambrian Moment. *The Economist*, special Report. January 18th 2014. About 540 - million years ago something amazing happened on the Earth: life forms began to multiply, leading to what is known as the "Cambrian explosion": until then sponges and other simple creatures had the planet largely to themselves, but within a few million years the animal kingdom became much more varied and interconnected.

future, a generational urgency other than a project for the future driven by knowledge, capacity and inclusiveness, which must work together in harmony.

Within a new political vision based on sharing knowledge, skills' impact may take various forms. The pervasive presence of the media, the wireless connection and the increasingly geolocalized social networks changes the way we communicate, think, feel, asses and decide. As a consequence, all areas of our lives are affected: work, investment, innovation, study, social cohesion and politics. Consequently, the expertise possessed by knowledge workers needs strengthening, and the same applies to knowledge leaders. It is not a matter of cognitive and rational practice, but rather of emotion, relationship and ethics other than the ability to understand, guide, change and mobilize diverse knowledge in order to deliver increasingly collective results. In order to provide leaders with the necessary tools to understand the dynamics they are about to implement, the socio-cultural know-how is a fundamental prerequisite.

Shirky (2010) describes the cognitive and leading force of crowdsourcing, the crowd that, by building common opinions and working together through the network produces a true "cloud politics": a widespread policy constantly enveloping us both as electors and decision-makers thus eliminating distances while reducing the pondering spaces. Political action and politicians' reaction merge in a short circuit that produces a virtuous participation on the one hand, and a vicious fragmentation of decision on the other hand. Not only do changes affect the economic and relational realm, but they are being, with growing pervasiveness, transferred to the physical realm, as regards physiognomy and physiology of the cities, intelligent. However, a smarter city is not the one whose traditional organization boasts the most intelligent and efficient technology, but the city that profoundly alters the development dynamics and revisits its housing and mobility patterns rethinking its metabolism through efficient urban cycles.

Increasing the infrastructural smartness is not sufficient, as cities ought to endeavour to increase the rate of collective intelligence, by supporting, via cloud communting, virtuous behaviour from the bottom and raising the profile of a new way to understand urbanism displaying its individual and collecting benefits. Smart communities are increasingly characterise by platforms for service whose value lies in the offered facilities considered useful by the users, which in turn translate them into additional services to other users. A sort of mutual complicity is therefore important between the platform and the value-adding users, which can be implemented provided the platform/user relationship is "transparent", "open" and "authentic" hence included in the new citizenship pact.

There are two words that best sum up the new virtuous relationship between information and open communities. The first is *Open Data* and is about the millions of available data that public administrations are networking in a challenging race to share data within a democratic knowledge process. The Open Data component is indispensable to open governance, in which governments are open to citizens both in terms of transparency and especially of direct participation in decision-making processes, promoting the use of ICT inasmuch as they accelerate the empowerment of communities, virtual in the first place and increasingly real today, other than generating renewed physical spaces fuelled by knowledge, sharing and inclusiveness.[2] One of the most interesting challenges is represented by *Open Gov* and how it can contribute to improve the "spatial based digital communities". The social movements set up online have increased inclusiveness and, in order to avoid infertile self-referentiality, they need to "tie up with the territory". According to Manuel Castells (2012) *online* and *offline* networks ought to be joined to obtain the new politics, that is the new "common city".

The other keyword is *Big Data*, the immense amount of data not only from government websites, but from social networks too, from blogs and specialistic websites that, if adequately designed, managed and interconnected, allow generating knowledge that could not be possibly obtained through traditional sources. Imagine planning the transport system of a city not only based on the service provider's information, but with the possibility to rely on users' feedbacks: their tweets, their complaints, thematic blogs, traffic information from the local police, data regarding ongoing or planned construction sites, information about planned strikes and demonstrations, the calendar of major events, citizens location trends and the list could be longer. All of these potential information must be handled within a city model allowing its use in terms of urban planning and design, otherwise they will just be "noise floor".

Open Data and Big Data management is not limited to the administrative sphere or to decision-making processes, but requires the traditional urban planning's cognitive model to be revised. It requires us not only to modify the protocols on which we base the plan's knowledge, but also to create new planning instruments. Hence, the first forms of *Open-source Urbanism* (Sassen 2011). We should therefore begin to outline it and experience its practices in order to identify the main application protocols. We find ourselves in a smarter dynamic and innovative context therefore. Above all, it is shared and open, and ought to be also more "senseable", aware and responsible. A proper Cloud Governance, not to be turned into a new mantra however: it ought to cooperate with leaderships and technocracies, with the directors and planners of the change, the actors in the transformation and the civil society to understand the extent to which the issue of openness

---

[2] The first large-scale experiment was carried out by the U.S. Government in 2009, launched by Barack Obama as a challenge based on the establishment of an informed and responsible community, able to be actively involved in the government decisions on the major current issues, such as environmental, social and health, then immediately extended to citizens participation in the urban-related choices by sharing the information possessed by experts and institutional decision-makers in an effective empowerment process. In Italy, the Open Governmental season opened in late 2011 following the www.dati.gov.it portal, in which the landscape managing process is gradually evolving towards increasingly open models, able to promote the setting up of a truly inclusive governance of territorial transformations. Recently, the Minister of Territorial Cohesion has launched the portal opencoesione.gov.it dedicated to the implementation of the 2007–2013 investments to allow citizens to assess whether the projects meet their expectations and whether the available resources are adequately employed, thus facilitating the reprogramming or activation of corrections and/or steps forward.

and transparency involve their organizations, be it businesses, institutions, communities or universities.

Planning in the cloud-based information age requires new maps, new sensors to detect obstacles, new tools for tracking the direction, but especially new eyes to not lose sight of the horizon.

8.2 Planning the City of Collective Intelligence

Everyday, we witness the application of new information and communication technology (ICT) to various areas of urban daily life: time management, traffic control, distribution and location of services, bureaucracy streamlining, dissemination of knowledge and communication, monitoring of the environment, not to mention the surrogate social and professional relationships of social networks. Technological innovation applied to production processes, remote home automation, explosion of mobile communications and the so-called "internet of things" in which many objects are interconnected and exchange information, are made available with ever-increasing pervasiveness for services' delivery and efficient management. In this way, urban services are enhanced and contribute to manage urban complexity, thus ensuring communications, relationships, dissemination of knowledge and culture benefiting citizens. We have entered the Open Data age, characterised by a daily increase in the number of databases and maps—often from institutional sources—available in digital format, not only intended for traditional institutional users or experts, but available for multiple uses, open to all potential users and for unpredictable uses. The diffusion of ICT throughout urban planning and urbanism processes marks the transition from a merely instrumental role in land management to a quality role in the management of transformations and in the participation, interpretation and orientation of new urban scenarios, as occurred with the *BMW Guggenheim Lab* held in New York, Berlin and Mumbai in search of the 100 Urban Trends for a better urban future, then presented at the Guggenheim Museum in 2013.

ICT advantages within planning and management processes are especially clear today, as maps, data and assessment models are increasingly becoming a common heritage: the integration of web and wiki technologies with GIS applications is a very fruitful way to improve the chances of constructive interaction between citizens, policy makers and the wise skills at stake within the urban planning processes. On cloud technologies, popular among professional and consumers alike, allow regular updates directly from the source through a steady integration of decentralized databases. Georeferenced systems are central to decision-making processes at local and regional level, facilitating decisions of institutional and entrepreneurial actors, for example by sharing land knowledge, encouraging fast-tracking of administrative procedures. Shared databases can encourage public–private partnerships and project financing by making data, information and feasibility studies available to technical offices or by ensuring multi-utilities contributions. Finally, the involvement of local partners or the international opening through an increasingly shared regulatory and planning power can thus be enhanced.

GIS tools are well-established in the field of urban and regional planning to improve the efficiency of planning and management processes. Today, a growing number of communities employ them to test—with increasingly widespread and interesting results, e-governance practices, which is the innovation of community services not only to provide a greater efficiency, but to increase urban rights, as new interface of the relationship between city and community, a planning method based on extended participation in the emerging forms of active citizenship and, finally, on the balance between cognitive opportunities offered by collaborative urban mapping. The spreading and integration of Web/GIS platforms in the public administrations and the popularity of Open Data not only contributes to improve the interpretation of resources and their better management, but encourages the establishment of a network of cities aimed at promoting local development, the enhancement of community and businesses services, by strengthening ties with networks of cities at international level. In addition, the cross-platform sharing of land use knowledge increases opportunities for new working activities, meaning opening up new spaces for higher education, lifelong learning and for the repositioning of broad professional categories of workers, especially young.

The development and spreading of new technologies in the field of urban and regional planning shall lead to the experimentation of new interfaces: representation and communication methods enhancing traditional systems to extend the application of geodata within the planning process. In this way, the planning process would change through new modalities to read and understand macro regions, strategic platforms, local systems and on-going socio-economic relations, revealing concealed links which would lead to a non-institutional redefinition of territorial aggregations.

The experimentation, carried out in several local realities, of GIS network projects, aimed at promoting networking of cities, is included within the broader challenge of promoting cloud governance as a new dimension of local development. Community is the sphere of ICT integration into urban policies where the communicative potential is best expressed, and added value is ensured, namely the combination of actors who, out of a common interest, interact within networks by carrying out transactions and exchanges, reporting problems and sharing solutions, developing projects and promoting actions aimed at increasing the added value. Land management as a system of interconnected sensors, together with interfaces and City Apps, for example, can encourage the setting up of virtual districts (in the fields of production, tourism, food, culture) based on cloud computing dedicated to SMEs with the goal of restoring the local system's competitive advantages, stimulating the region's integrated development by linking businesses with other global enterprises networks. The local districts philosophy will lead city networks to compete in the global market as local network systems, employing three important competitive resources: geolocalised information, digital connections and citizen networks.

The combination of technological innovation and urban planning, however, is not only instrumental and determines changes within the community and its territory too. The "Third Industrial Revolution" and the gradual implementation of e-Society has made it possible to delegate an increasing number of physical and intellectual tasks, even very sophisticated, to technology. In fact, the goods and ideas produced are increasingly less tied to a scheduled place and time, in terms of quality and quantity; the workplace is no longer an independent variable and time is no longer rigidly synchronized, especially as far as the intellectual work is concerned.

The spreading of sensors, electronic networks and urban life apps has created a proper urban cyber-physical space, consisting of the constant interaction between physical components and digital networks, tangible actions and intangible feedback. "We are at the onset of a hybrid dimension between the digital and material world, where the Internet is invading the physical space"—claim Ratti and Sassen (2009)—by identifying it, making it attractive and setting it up for social uses, which are expected to gather the citizens in smart places connected to the network and providing services. We are witnessing the evolution of the cyber cafés: mobile connection disengages the user from a fixed location and brings him back into the city, parks, waterfront and squares allowing him to communicate and interact, learn and point out, know and judge. The dematerialization of technology and its on cloud and mobile spreading allows citizens to "re-materialize" themselves in the city.

The associated research and planning efforts provide a complex framework ranging from the debate on the effects of technological innovation to the analysis of changes occurring within the location of urban settlements and productive activities or in the structure of transport networks and the related infrastructures. Social and economic transformations, brought about by information and communication technologies, lead planners to investigate changes in resources' exploitation and the evolution of certain qualitative aspects of territorial organization (especially with regard to education and leisure structures). Finally, opportunities offered by the Information Society provide planning and land management with new tools, resources, real and virtual subjects. Consider, for example, the growing role of civic networks in the processes of communication and participation in the plan. ICT applied to collective decisions to monitor the effectiveness of the actions, understanding processes and promoting partnerships returns citizens their leading role in the civil society, thus contributing to an adequate distribution of the powers of the plan.

Introducing technologies, protocols and communication digital devices in the urban organism is not only an opportunity for innovation and participatory cognitive processes but should provide an opportunity to redefine development, competitiveness and cohesion in order to give the city a swing power able to overcome the tsunami of the crisis (Siemens-Cittalia 2012). The future is challenged to focus on Smart Cities, as long as they manage to gather skills, generate creativity as innovation incubators empowering communities, in addition to being drivers of competitiveness. Otherwise they risk becoming just "cemeteries of obsolete machines" as fears Saskia Sassen (2011), who is suggesting hacking the city to facilitate its transformation through informal actions performed by the citizens' collective intelligence.

Therefore, a smart city is not only a more intelligent, technologically managed and efficient city but more skilled, fair and equal (Bullard 2007) which profoundly innovates its sources of knowledge, dialectic capacity, development dynamics and revises its settlement patterns: a more "ingenious" city" (Granelli 2012). The smart city passes from being reactive to proactive by effectively using a better and broader information flow. It invests in people—men and women before human capital—enhancing their ability to empower the social capital, strengthening participation processes, extending education and spreading culture by improving the new mobile communications infrastructures (Campbell 2012). It focuses simultaneously on software and hardware, to ensure a higher quality of life for all citizens with an accountable resources management through cooperative governance practices.

They are defined *Smart and Creative Cities* (Carta 2014) as they will have to be able to innovate high-impact areas: planning, urban design and land management, energy production-distribution-consumption cycle, transport of goods, development of mobility for people and freight, buildings energy efficiency and active participation. Complex realms, involving several actors such as education, health, waste as well as the enhancement and use of the cultural heritage and tourist attractiveness will have to be innovated. However, cities can not limit themselves to their infrastructures, but shall contribute to increase the rate of "collective intelligence". Moreover, a city that aspires to be skilled and resourceful needs to show solidarity too supporting, through cloud commuting, bottom-up virtuous behaviours from below by emphasising the individual and collective benefits of open urbanism.[3]

Therefore, urban smartness should be more than just an adjective that applies to traditional governance, planning and management methods, but rather a challenge to gain tacit skills and generate new knowledge, creativity and innovation. The sustainable development of a smart city is based on the comprehensive rethinking of its metabolism focused on waste reduction, separation and collection and consequent economic exploitation, on the efficiency of urban agriculture and the drastic reduction of greenhouse gas emissions. To this end, the reorganization of private car traffic and the optimization of industrial emissions is of pivotal importance. The improvement of the construction industry and the housing market through a real innovation of the buildings in terms of structural efficiency, reorganization of public lighting and better management of urban green areas are the

---

[3] In Helsinki open data, living lab and crowdsourcing are now daily items on the agenda, accelerated by the *Forum Virium Helsinki 's Smart City Project Area* with the purpose of making the metropolitan region of the Finnish capital a good practice as for the provision of digital services within urban regeneration processes, starting from *Arabianranta*, the new creative district of the city.

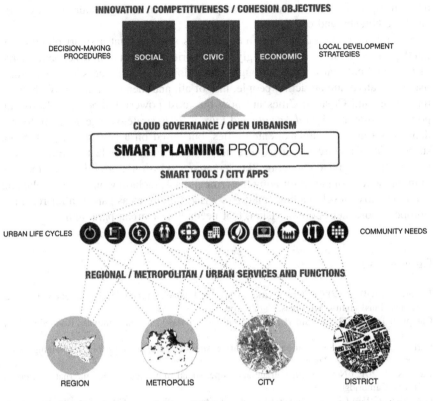

Fig. 8.1 Smart planning protocol for rethinking urban planning as an intelligent integrator between cloud governance and smart tools (© SmartPlanning Lab, Palermo, 2014)

challenges that lie ahead for the necessary reduction of the urban impact on the environment.[4]

If it is true that the traditional density, centrality and urban form European model is now "obsolete in describing the contemporary city; necessarily, the open space between buildings ought to be reconsidered, re-setting infrastructure systems adopting the most appropriate scale, which is the geographical and landscape one" (Waldheim 2009). A smart and creative city shall therefore adopt a broad and panoptic view targeting its development to the implementation of citizens' quality of life, investing in the quality of services and public space, on safety, innovative and flexible lifestyle models; but also adopting a settlement pattern whose

---

[4] The Strategic Plan "Smart Cities Copenhagen Vision" turns the entire city into a sustainable laboratory for mobility and energy, primarily focusing on the port district of Nordhavn. Even Stockholm, *European Green Capital 2010*, aims to become independent from fossil fuels by 2050; it has already reduced the 1990 emissions level by 25 %, notwithstanding a significant rise in population and is investing in new technologies within service management applied to the eco-district of Hammarby Sjöstad.

relationship with urban land and landscape is based on saving resources, recycling, energy efficiency and creativity.

Several international analyses and rankings show us that a smart planning of intelligent cities must be increasingly culture-oriented, with policies enhancing the identity and cultural heritage through internationalization processes, or by establishing creative hub attracting people, first of all, and then economies. We believe that Smart and Creative Cities are now the most powerful support of European policies within the *Digital Agenda* and *Europa 2020* strategy to generate no longer debit-based and consumer-oriented cities, but based on a new social pact, more sustainable. They are the DNA of a new urban organism able to rethink development (Fig. 8.1) and encourage the "creative explosion", provided they manage to implement strategies improving the critical mass, including on the symbolic and communicative level, leading smartness-based initiatives as part of a far-reaching European post-metropolitan vision, that needs a new urban paradigm.

References

Bullard RD (ed) (2007) Growing smarter: achieving livable communities, environmental justice, and regional equity. Mit Press, Cambridge

Campbell T (2012) Beyond smart cities: how cities Network, learn and innovate. Routledge, New York

Carta M (2014) Reimagining urbanism. Creative, smart and Green Cities for the changing times. ListLab, Barcelona-Trento

Castells M (2012) Networks of outrage and hope: social movements in the internet age. Polity Press, Cambridge

Granelli A (2012) Città intelligenti? Per una via italiana alle Smart Cities. Sossella, Bologna

Ratti C, Sassen S (2009) Le megacittà iperconnesse. Aspenia, 44

Sassen S (2011) Open-source urbanism. The new city reader: a newspaper of public space, n. 14, Jan

Shirky C (2010) Cognitive surplus: creativity and generosity in a connected age. Penguin, London

Siemens-Cittalia (2012) EfficienCITIES. Città-modello per lo sviluppo del paese. Milano, Siemens

Waldheim C (2009) The other 56. In: Krieger A, Saunders WS (eds) Urban design. University of Minnesota Press, Minneapolis